D0938014

Enterprise Resource Planning Systems

Systems, Life Cycle, Electronic Commerce, and Risk

Enterprise resource planning (ERP) systems are powerful software packages that enable businesses to integrate a variety of disparate functions. In particular, ERP systems can provide the foundation for a wide range of e-commerce–based processes, including web-based ordering and order tracing, inventory management, and built-to-order goods. This book examines the pros and cons of ERP systems, explains how they work, and highlights their role at the heart of e-commerce.

The author begins by explaining the background of ERP systems and goes on to discuss specific systems, such as SAP, and their capabilities. He then focuses on the ERP life cycle, from the decision on whether or not to adopt an ERP system to the time when the system goes "live." After covering the use of ERP in e-commerce, he concludes by discussing the risks associated with the adoption of ERP systems.

The book contains several detailed case studies and will be an invaluable guide to managers and consultants working with ERP systems. It also will be a useful reference for MBA students taking courses in information systems management.

Daniel O'Leary received his Ph.D. from Case Western Reserve University and his MBA from the University of Michigan. He is a Professor in the Marshall School of Business at the University of Southern California. He has published over 120 papers in a variety of computer science, information systems, and management science journals.

Enterprise Resource Planning Systems

Systems, Life Cycle, Electronic Commerce, and Risk

DANIEL E. O'LEARY

University of Southern California

CAMBRIDGE
UNIVERSITY PRESS

PUBLISHED BY THE PRESS SYNDICATE OF THE UNIVERSITY OF CAMBRIDGE
The Pitt Building, Trumpington Street, Cambridge, United Kingdom

CAMBRIDGE UNIVERSITY PRESS
The Edinburgh Building, Cambridge CB2 2RU, UK   www.cup.cam.ac.uk
40 West 20th Street, New York, NY 10011-4211, USA   www.cup.org
10 Stamford Road, Oakleigh, Melbourne 3166, Australia
Ruiz de Alarcón 13, 28014 Madrid, Spain

First published 2000

Printed in the United States of America

Typeface Times 10/13 pt.   *System* AMS-T$_E$X   [FH]

A catalog record for this book is available from the British Library

Library of Congress Cataloging in Publication Data
O'Leary, Daniel Edmund.
Enterprise resource planning systems : systems, life cycle,
electronic commerce, and risk / Daniel E. O'Leary.
p.   cm.
ISBN 0-521-79152-9 (hb)
1. Production management.   2. Management information systems.
3. Business planning.   4. Electronic commerce.   I. Title.
TS155.O385   2000
658.8′4 – dc21   00-027554

ISBN 0 521 79152 9 hardback

Contents

v

PART ONE

INTRODUCTION AND BACKGROUND

1

Introduction

This chapter initiates our dialogue into enterprise resource planning (ERP) systems, focusing on the following questions.

- Why investigate ERP systems?
- How does ERP create value?
- What is the purpose and scope of this book?
- What is the outline of this book?

Why Investigate Enterprise Resource Planning Systems?

Enterprise resource planning systems are a corporate marvel, with a huge impact on both the business and information technology worlds, including each of the following dimensions:

- ERP affects most major corporations in the world;
- ERP affects many SMEs (small and medium enterprises);
- ERP affects competitors' behavior;
- ERP affects business partner requirements;
- ERP has changed the nature of consulting firms;
- ERP provides one of the primary tools for reengineering;
- ERP has diffused many "best practices";
- ERP gave client server computing its first enterprise product;
- ERP has changed the nature of the information systems function;
- ERP has changed the nature of jobs in all functional areas;
- ERP cost is high;
- ERP has experienced huge market growth.

ERP Affects Most Major Corporations in the World (Bowley 1998). A single ERP system (SAP's R/3) is used by more than 60% of the multinational firms. Further, according to Arthur D. Little's global strategy leader, an ERP company (SAP) "is conquering the world. Almost every important company is more or less in its hands."

ERP Affects Many SMEs (Foley and Stein 1997). The impact of ERP is not limited to large firms. In 1995, SAP generated 90% of their revenues from large global companies, but by 1997 SAP expected 50% of its revenues to come from SMEs (small and medium enterprises). Roughly 35% of SAP's 1997 customers had revenues of under $200 million.

ERP Affects Competitors' Behavior. On June 24, 1996, Oracle's Application Division announced that "[s]everal companies went live with their Oracle Applications implementations during the quarter, including Silicon Graphics, Inc. and Quantum Corporation, both of whom successfully deployed large-scale implementations." In addition, at the same time, Oracle's Application Division announced that "among the customers added that quarter included ... Western Digital ...." Western Digital was a direct competitor of Quantum.

When one corporation adopts ERP, should its competitors do the same? If the software provides a competitive advantage and/or would create value, then the answer is probably Yes. But which software should they choose and who should implement it for them? We might expect that, if one company successfully implements ERP for competitive advantage, then the same software and consulting team would be chosen to implement ERP for its competitors – after all, who would have greater experience with the industry? Yet how would the firm that originally implemented ERP react to such an occurrence?

ERP Affects Business Partner Requirements. Generally, adopting ERP makes firms more "information agile." Those firms can better process information and integrate it into their business procedures and decision making. As a result, business partners need to adapt to the changes that will occur in ERP-adopting organizations. For example, as ERP-adopting firms operate in real time, they will expect the same of their partners. Further, ERP-adopting firms may begin to integrate ERP systems along the supply chain, potentially pushing ERP to other parts of the supply chain, which in turn are likely to include their partners.

ERP Has Changed the Nature of the Largest Consulting Firms. Enterprise resource planning systems have been critically important to the growth of consulting among the "Big 6" (recently the "Big 5") and other professional service firms. According to *Public Accounting Report* (1998), services involving ERP packages generate one third to one half of the total consulting revenue at national professional services firms.

ERP Provides One of the Primary Tools for Reengineering. In 1990, Hammer's highly influential article on reengineering sparked the corporate world's interest in obliterating existing processes. Unfortunately, after things were obliterated many firms had no idea what to replace them with. Enterprise resource planning provides perhaps the primary tool for guiding those efforts, so much so that Gendron (1996) called ERP (particularly SAP's R/3) the "electronic embodiment" of reengineering and Hammer (1997) commented that "SAP equals forced reengineering."

ERP Has Diffused Many Best Practices. Enterprise resource planning systems are based on so-called best practices – the best ways of doing processes. SAP's R/3 incorporates over a thousand of them! What this means is that any firm that installs R/3 has access to a wide range of best practices. Furthermore, new business practices are being added all the time. As new best practices are found and embedded in particular applications, they become available for inclusion in new versions of R/3; as they become available, other firms install them. Hence there is a cycle of finding best practices, building them into the software, and diffusing them out to new users.

Firms should therefore ask the following questions.

- What business processes give us competitive advantage?
- What new processes would give us competitive advantage?
- How do we know that there is something unique in those processes?
- Is this process worth more or less than moving to a set of processes that are widely available?
- What would be the cost of this best practice diffusing to my competition?

ERP Gave Client Servers Their First Enterprise Product. In the early 1990s, client server computing was an available technology that offered many advantages over existing mainframe solutions. Unfortunately, there was limited software available to exploit the advantages. Enterprise resource planning changed all that when it became one of the first dominant corporate applications of client server computing.

ERP Has Changed the Nature of the Information Systems Function. Historically, the job of the information systems function was primarily one of designing, developing, and implementing software. Now, with ERP systems, the design and development functions are being outsourced. Enterprise resource planning systems are replacing major portions of most firms' software needs. This changes the basic nature of the information systems function from one where systems analysts and programmers are needed to one where knowledge of existing software packages is critical.

Not only have needs changed, but personnel have become more mobile. Historically, information systems personnel would have knowledge only of firm-specific legacy applications. With ERP software that changes, knowledge can be used at more than one firm. Knowledge of almost any ERP package is useful not just in one organization, it is useful around the world. Thus, as the use of ERP package software grows, there is more mobility among personnel in information systems than has ever been seen.

In addition, this mobility is changing the consulting business that supports ERP package software. Consultants armed with knowledge about such a package can now take that knowledge from one firm to another. The consultant actually becomes more and more valuable with each new implementation of the software.

ERP Has Changed the Nature of Jobs in All Functional Areas. Enterprise resource planning has changed the nature of jobs in functional areas, such as manufacturing. As noted by Corcoran (1998),

> IT [information technology] professionals in manufacturing say ERP systems are blurring the lines between IT and users. There is a huge demand for users or line-of-business personnel who also have professional level IT skills. But traditional IT types who know only about technology and nothing about the business are not needed now as they once were. "Understanding the business is probably the most critical [aspect,]" says Joan Cox, CIO at the space and strategic missiles sector of Lockheed Martin in Bethesda, MD. "Its more important to understand how you want things to flow through your factory than [to have] the skill of programming – except for the few places where SAP doesn't do what's needed so you need coders."

ERP Cost Is High. According to the META group, the average cost of ownership for an ERP implementation is $15 million, typically at a cost of $53,320 per user. These estimates include software, hardware, professional services, and internal staff costs for the full implementation, plus two years of post-implementation support. As noted by Escalle, Cotteleer, and Austin (1999), ERP costs can run as high as two to three percent of revenues.

ERP Has Experienced Huge Market Growth. According to Frye (1994), in 1993 – the early days of client servers – five vendors accounted for 74% of the client server ERP software: Oracle, $88 million; SAP America, $71 million; D&B Software, $30 million; IMRS, $30 million; and Computron, $17 million. The entire market was $319 million. In 1998 the license/maintenance revenue market was $17.2 billion, and in 2000 the market is expected to be $24.3 billion (PricewaterhouseCoopers 1999). The market growth in ERP has been huge.

How Does ERP Create Value?

Historically, legacy information systems have been functionally based and not integrated across multiple locations or functional areas. The same information was captured multiple times, in multiple places, and was not available in real time. Jobs and processes were narrowly defined in concert with the division of labor and the industrial revolution. As a result, some information never made it out of different pockets of the corporation. Processes and job definitions saw to it that information remained a local good. When information did "go global," often there were different informational reports of the same events. Thus, there were information asymmetries between the different local and functional groups and top management.

Enterprise resource planning systems provide firms with transaction processing models that are integrated with other activities of the firm, such as production planning and human resources. By implementing standard enterprise processes and a single database that spans the range of enterprise activities and locations, ERP systems provide integration across multiple locations and functional areas. As a result, ERP systems have led to improved decision-making capabilities that manifest themselves in a wide range of metrics, such as decreased inventory (raw materials, in-process and finished goods), personnel reductions, speeding up the financial close process, and others. Thus, ERP can be used to help firms create value. In particular, ERP facilitates value creation by changing the basic nature of organizations in a number of different ways.

ERP Integrates Firm Activities

As noted by Hammer (1997), "[i]ntegration is the defining characteristic of SAP." Enterprise resource planning processes are cross-functional, forcing the firm out of traditional, functional, and locational silos. In addition, an organization's different business processes are often integrated with each other. Further, data that were formerly resident on different heterogeneous systems are now integrated into a single system.

ERPs Employ Use of "Best Practices"

Enterprise resource planning systems have integrated within them a thousand or more best practice business processes. Those best practices can be used to improve the way that firms do business. Choice and implementation of an ERP requires implementation of such best practices.

ERP Enables Organizational Standardization

Enterprise resource planning systems permit organizational standardization across different locations. As a result, those locations with substandard processes can be brought in line with other, more efficient processes. Moreover, the firm can show a single image to the outside world. Rather than receiving different documents when a firm deals with different branches or plants, a single common view can be presented to the world, one that puts forth the best image.

ERP Eliminates Information Asymmetries

Enterprise resource planning systems put all the information into the same underlying database, eliminating many information asymmetries. This has a number of implications. First, it allows increased control. As noted in Brownlee (1996, p. R17) by one of the users, "[i]f you don't do your job, I can see that something hasn't been done." Second, it opens up access to information to those who need it, ideally providing improved decision-making information. Third, information is lost as a bargaining chip, since information is now available both up and down the organization. Fourth, it can "flatten" an organization: because information is widely available, there is no need for non–value-adding workers whose primary activity is to prepare information for upward or downward dissemination.

ERP Provides On-Line and Real-Time Information

In legacy systems, much information is captured on paper and then passed to another part of the organization, where it is either repackaged (typically aggregated) or put into a computer-based format. With ERP systems, much information is gathered at the source and placed directly into the computer. As a result, information is available on-line to others, and in real time.

ERP Allows Simultaneous Access to the Same Data for Planning and Control

Enterprise resource planning uses a single database, where most information is entered once and only once. Since the data is available on-line and in real time, virtually all organizational users have access to the same information for planning and control purposes. This can facilitate more consistent planning and control, in contrast to legacy systems.

ERP Facilitates Intra-Organization Communication and Collaboration

Enterprise resource planning also facilitates intra-organization (between different functions and locations) communication and collaboration. The existence of interlocking processes brings functions and locations into communication and forces collaboration. The standardization of processes also facilitates collaboration, since there are fewer conflicts between the processes. Further, the single database facilitates communication by providing each location and function with the information that they need.

ERP Facilitates Inter-Organization Communication and Collaboration

The ERP system provides the information backbone for communication and collaboration with other organizations. Increasingly, firms are opening up their databases to partners to facilitate procurement and other functions. In order for such an arrangement to work there needs to be a single repository to which partners can go; ERP can be used to facilitate such exchanges.

What Is the Purpose and Scope of This Book?

The purpose of this book is to examine some of the most important and interesting issues, cases, and ideas associated with ERP systems. However, this book is not an ERP encyclopedia. It focuses on those issues that are critical to consultants and managers. Also, it focuses on those notions that are more unique to ERP rather than to software in general. For example, although project management is critical to any ERP implementation, it remains project management, and a great deal of material is available on information systems project management in other sources. The book does not focus on the hands-on issues, since those issues differ from software to software and information is available from a number of other sources. For example, a detailed analysis of SAP is available in ASAP (1996) and Curran and Keller (1998). This book rarely focuses on details associated with a particular ERP system, although specific systems are sometimes used to illustrate certain concepts.

What Is the Outline of This Book?

This book focuses on five primary aspects of ERP systems:

  (1) background (Chapter 2),
  (2) systems and their capabilities (Chapters 3–6),
  (3) ERP system life cycle (Chapters 7–13),
  (4) electronic commerce (Chapter 14), and
  (5) risk (Chapter 15).

Background

A wide range of information and other technologies are necessary to drive ERP systems, including client server computing, networks, relational database systems (and data warehouses), software concepts (including package software and legacy software), requirements analysis concepts (e.g., "as is" modeling), and reengineering.

Systems and Capabilities

Analysis of ERP systems and capabilities starts with the ERP vendors and partners, reviews some of the modules available in two ERP systems, and discusses issues such as using ERP software from a single vendor or a "best of breed" approach. In addition, the models and processes at the base of ERP applications are reviewed, along with a short summary of how ERP systems work. Then a detailed analysis of ERP data input and output is made in order to determine potential sources of ERP costs and benefits. Finally, the advantages and disadvantages of "clean slate" reengineering versus ERP technology enabled reengineering are discussed.

ERP Life-Cycle Issues

The chapters in Part Three of this book incorporate a general life-cycle model of the process that a firm goes through with ERP systems. This cycle may be broken down as follows.

- Deciding to Go ERP
- Choosing an ERP System
- Designing ERP Systems
  - Should Business Processes or ERP Software Be Changed?
  - Choosing Standard Models, Artifacts, and Processes
- Implementing ERP: Big Bang versus Phased
- After Going Live
- Training (an issue of concern *throughout* the entire life cycle)

Electronic Commerce

Enterprise resource planning provides information backbone that can provide a basis for building electronic commerce applications. Ultimately, ERP systems must integrate with other systems, or ERP vendors must generate their own solutions to electronic commerce. In either case, ERP systems can facilitate electronic commerce.

Risk

This introduction has focused on the positive side of ERP systems. Yet where there are huge opportunities for growth and value creation, there are also huge opportunities for risk. Our analysis presents a model that is based on identifying risk throughout the life cycle.

Materials in This Book

Chapters

Each chapter addresses a relatively independent chunk of ERP material. In general, however, the material builds as the book moves from chapter to chapter. Throughout, a wide range of real-world examples are used to illustrate or generate major points about ERP systems. Those examples come from the literature as well as from interviews with a number of companies.

Appendices

Appendices take three different forms. First, there are "long cases." Geneva Steel (Appendix 3-1) provides insight into expectations for ERP systems and addresses some cultural issues of change. Microsoft (Appendix 9-1) provides an investigation into the different organizational entities involved in the choice of an ERP system and allows analysis of the issue of whether to change processes or software. Quantum (Appendices 11-1–11-3) provides insight into the implementation process.

Second, there are "short cases." Quantum (Appendix 5-1) allows analysis of virtual data warehouses and how they interface with ERP systems. Chesapeake Display and Packaging (Appendix 8-1) provides a real-world example of a firm's choice process between different ERP systems. Appendix 8-2 summarizes one inquiry that I received from a CFO regarding his firm's choice of an ERP system. A case study of XYZ Company (Appendix 12-1) provides insights into some of the concerns that a medium-sized firm had regarding evaluation of its ERP system. An interview with Les Porter (Appendix 14-1) provides some insights into an emerging form of Internet-enabling ERP, using J.D. Edwards as an example.

Third, the appendices provide more detail on ERP information. "In-House or Outsourced" (Appendix 7-1) provides some drill down on an increasingly important way of going ERP. Deloitte Consulting's post-implementation checklist (Appendix 12-2) provides a quick summary of some key post-implementation information.

References

ASAP [World Consultancy] (1996). *Using SAP R/3.* Indianapolis, IN: Que.

Bowley, G. (1998). "Silicon Valley's Transplanted Sapling." *Financial Times,* March 27.

Brownlee, L. (1996). "Overhaul." *Wall Street Journal,* November 18, pp. R12, R17.

Corcoran, C. (1998). "ERP is Changing Manufacturing Jobs." *InfoWorld,* July 13.

Curran, T., and Keller, G. (1998). *SAP R/3 Business Blueprint.* Upper Saddle River, NJ: Prentice-Hall.

Escalle, C., Cotteleer, M., and Austin, R. (1999). "Enterprise Resource Planning (ERP)." Report no. 9-699-020, Harvard Business School, Cambridge, MA.

Foley, J., and Stein, T. (1997). "Oracle Aims at Applications Midmarket." *Information Week,* July 7, p. 30.

Frye, C. (1994). "With Financial Apps, DBMS Support Often Drives the Sale." *Software Magazine,* June, pp. 55–7.

Gendron, M. (1996). "Learning to Live with the Electronic Embodiment of Reengineering." *Harvard Management Update,* November, pp. 3–4.

Hammer, M. (1990). "Reengineering Work: Don't Automate, Obliterate." *Harvard Business Review,* July/August, pp. 104–12.

Hammer, M. (1997). "Reengineering, SAP and Business Processes." Unpublished presentation given at SAPPHIRE (Orlando, FL), August.

PricewaterhouseCoopers (1999). *Technology Forecast, 1999.* Palo Alto, CA: PricewaterhouseCoopers.

Public Accounting Report (1998). "Big Six Dominate Systems Integration Market." July 31, p. 4.

2

Systems and Technology Background

The purpose of this chapter is to provide some brief background information on systems and technology to facilitate understanding of ERP issues. In order to do that, some background on computing, networking, databases, software, software choice, and reengineering is provided. Within each category, the background information is applied to some related ERP (in particular, SAP) concepts in order to tie the background material directly to ERP.

Computing

Historically, enterprise computing was heavily mainframe based. In roughly 1993 (see e.g. Frye 1994), client server versions of enterprise computing applications began to be more available.

Mainframe Computing

In a mainframe computing environment, all the computing is done on a single computer (box). Typically, this is handled by allowing users to share (e.g., via time sharing) the computing resource – the mainframe computer. The user would typically be at a terminal that had no computing capabilities (or at a personal computer that emulated a terminal).

Client Server Computing

Over time, the user began to have increasing computing capabilities locally. As a result, computing began to shift some of the processing to the user's computing facilities. This ultimately resulted in what is now called "client server" computing, where client and server are linked so that the computing and storage can be distributed between the client and the server. The term "client" refers to the user's computer, while the "server" refers to some other (e.g., centrally located)

Client/Server Configurations

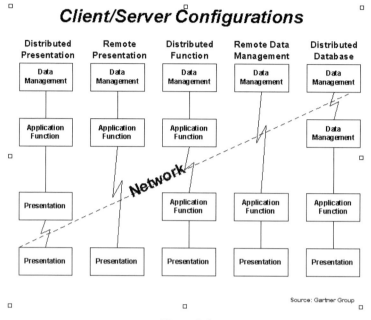

Figure 2.1.

computing source that provides computing resources, software, or data. A spectrum of different types of client server computing is summarized in Figure 2.1.

The terms "thin" or "fat" client generally refer to the processing capability of the client. Thin and fat are on a continuum that seems to keep changing as technology capabilities increase. An alternative set of terms, possibly with different connotations, are "weak" and "powerful." Typically, thin (or weak) refers to a client with relatively few computing capabilities. Most of the computing in a thin-client environment would done at the server. Similarly, fat (or powerful) refers to a client with greater computing capabilities.

In three-tier client server computing, which is often used in enterprise computing, there are three servers (or sets of servers) with different tasks. Servers handle the application software, other servers the database software, and still others the user interface. There can also be associated servers that provide support services to the application servers. These support services can include dialog management, gateway services, and other activities.

SAP and Client Server

SAP's original product (R/2, developed for a mainframe environment) was introduced in 1974. SAP's client server software, R/3, was introduced in 1992 in Europe and in 1993 in North America.

R/3 can be configured in a number of ways (ASAP 1996). At one end of the spectrum, a central system can be used to provide presentation, applications, and database services. At the other end of the spectrum, larger ERPs employ a three-tiered client server arrangement with separate servers for presentation, applications, and database (ASAP 1996; Juergens 1999).

Networks

In a client server setting, there is a network between clients and servers; this network may be a local area network (LAN) or across the Internet. Network capabilities, standards, and security are critical to the success of any system. This section provides a very brief overview of some network concepts that will be used later in the book.

LANs, Intranets, and Extranets

Local area networks are networks used to link computers together over a relatively small geographical area, such as within a building. Wide area networks (WANs) link together computers over a large area. *Intranets* typically are WANs that are only for use by a specific corporation; *extranets* typically are WANs that are for use by a specific corporation and its partners.

Bandwidth, Standards, and Security

A network must be able to accommodate the requirements placed upon it by the applications. *Bandwidth* is a network's transmission capacity. The greater the bandwidth, the greater the network's capacity.

Standard protocols allow the transmission of information in a common form, facilitating interaction between components. Perhaps the most widespread network standards are those associated with the Internet: transmission control protocol (TCP) and Internet protocol (IP).

Two of the many approaches to securing information are firewalls and encryption. Firewalls provide a measure of security in networks. Firewalls enforce a site's security by controlling the flow of traffic. Typically, firewalls are placed between the corporation's network and an external network (e.g., the Internet). Encryption provides some security for information placed on open networks. In particular, encryption converts readable information into a generally unreadable format. Information security can be obtained by encrypting information so that only those who can decode the encrypted information will have access to it.

SAP's Communications

If an ERP has a stand-alone network then the network's bandwidth must be large enough to accommodate the transaction flow from the ERP. Otherwise, the network must accommodate the ERP transactions and any other application activity put on the same network.

Communications within ERP typically use standard communication protocols. For example, SAP's R/3 employs TCP/IP to handle communications between client server configurations (ASAP 1996).

Two Security Approaches in SAP

SAP offers a number of approaches to security. This section discusses SAProuter and encryption (see Juergens 1999).

SAProuter is software that provides an extra layer of security. SAProuter transports SAP connections across a firewall. The SAProuter does not replace the requirement for a firewall but rather is used in conjunction with firewalls. The SAProuter is used to either explicitly deny or allow connections from a particular machine or some subset. SAProuters can be used to connect trading-partner SAP systems.

SAP communications across the network are not encrypted. However, for those organizations or settings that require more secure data, SAP provides the ability to encrypt certain communications.

Databases

Most enterprise software, such as ERP systems, employ relational databases in order to capture and make available the enterprise data to all modules of the software. However, some enterprise software still is based on "flat file" databases that are not directly related to each other.

Flat File Databases

Flat files are two-dimensional (rows and columns) files. Flat files are easy to use but are limited in their ability to model enterprise events. Flat file databases typically are used in software aimed at a specific application. For example, a sales system is likely to have information that relates to salespersons, as seen in Figure 2.2. Flat file databases are not directly related to other flat files.

The problem with flat files is that there may be substantial redundancy. For example, consider a flat file designed to capture data that relates to the sales of goods. As part of the database we would want to capture who made the sale,

Salesperson #	Name	Address	...

Figure 2.2. Flat File

their address and social security number, and so forth. However, if that same person makes another sale then the database would capture all that same information again, resulting in substantial redundancy. One approach to eliminating some of that redundancy is to use relational databases, where a single reference number that links information can be used to capture all the information about who made the sale.

Relational Databases

A relational database is a set of database tables that are related or linked to each other (see Figure 2.3). For example, rather than capturing a complete set of customer (salesperson) information for each sales order, only the customer (salesperson) number needs to be captured. Then, if more detailed information is needed, we can look it up in the related table by using the customer (salesperson) number.

Relational structures are quite robust and can handle a wide range of enterprise events. In Figure 2.4, for example, more detailed sales information is captured about the items and amounts of the goods in the sales order. Using the relational approach, a new line is introduced for each different inventory item sold. As a result, there is no need to plan ahead with regard to how many items are expected to be in a sale and other related issues.

Data Warehouses

A data warehouse is "a single place located across a corporation's networks where any user can get the latest data, efficiently organized" (Radding 1995,

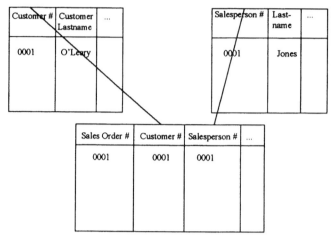

Figure 2.3. Sales Order Information

Sales Order #	Inventory #	Amount ...		Inventory #	Name	Description ...
0001	0001	20		0001	Roter	....
0001	0002	30		0002	Top	....
0002	0002	50				
0003	...					

Figure 2.4. Capture of a Sales Event

p. 57). Data warehouses are large repositories of data that have been optimized to permit rapid response to user queries. Data warehouses typically include multiple years of transaction data so that users can make decisions based on the trends revealed by data analysis.

Software

Ultimately, ERP systems are software, and any discussion of ERP software must cover a range of software issues that include operating systems, legacy systems, package software, database management systems, and the existence of different software versions.

Operating Systems

Enterprise resource planning systems have been designed to run under multiple operating systems, including UNIX-based systems and Windows NT. Initially, virtually all large ERP implementations were done in UNIX. However, as noted in ASAP (1996) for SAP's R/3, each of the operating systems for the presentation, application, and database can be different.

Legacy Software

Legacy software refers to software that preceded the software that is now being implemented or considered for implementation. Legacy software typically has been developed by and for the specific firm. Legacy software is often mainframe software. The organizational requirements for legacy software include a staff that can do systems analysis, design, and implementation of the software. Typically, there is substantial maintenance cost of a legacy system as it is updated to meet emerging organizational needs.

Package Software

Package software is software built for use in multiple organizations. Perhaps the best-known package software has been developed by Microsoft. At the individual PC (personal computer) level there is now a wide range of package software, including Microsoft Word and Microsoft Excel. Recently, the package software trend has encompassed enterprise computing as package software, so ERP systems are now among the software referred to as package software.

Package software is built with a large number of capabilities. Many of the capabilities are not actually used, as the individual company chooses from among the portfolio of capabilities available within the software. One approach to setting the capabilities to meet the needs of a specific company is to set parameters in the software to choose among the different capabilities. Access to the capabilities does not necessarily require reprogramming the system. Instead, this is often done by using "switch setting" – the equivalent of software switches turning options on and off are used to make choices among the capabilities.

Ultimately, package software means abandoning the internal development of software and instead "outsourcing" its development. As a result, "[b]uying an ERP suite means more than buying software. You buy that company's view"

(Langdoc 1998). Choosing ERP means that you delegate much of your systems maintenance and evolution to the ERP vendor. As a result, an important variable in choosing software can be the vision of the package vendor.

Database Management System

A database management system (DBMS) is the software designed to facilitate use of particular database structures – for example, relational database systems. Database management systems are useful not only because they have data in them but also because they allow access to the data. Information is generated from databases by using reports and queries. Database management systems therefore include data query commands (which enable the user to ask questions about the data) and report generators (which allow development of reports about the data).

Typically, ERP systems are developed so that they can use any of a number of database management systems. For example, SAP's R/3 supports database management systems from Oracle, Microsoft's SQL, Sybase, IBM's DB2, and Informix.

In some cases, firms constrain their choice of ERP systems to those that are compatible with their existing database system. If this is true then we can anticipate that ERP vendors would be most interested in having their products support the dominant database management systems. In 1997, Oracle had a 39.2% market share, IBM a 13.3% share, Informix 5.6%, Sybase 5.4%, and Microsoft a 4.5% share of the database management systems market (Markoff 1998).

On the other hand, the choice of the database system is dependent on the operating system. However, as noted by Juergens (1999), after implementation it is virtually impossible to change the database management system. Similarly, the ability to upgrade the ERP software will depend on the new version's compatibility with the specific DBMS.

Historically, legacy software was treated independently of the database management systems. However, in ERP systems such as SAP (Juergens 1999), the database management system is integrated into SAP via the "basis" component.

Versions of Software

Package software comes in different versions, where new versions typically have greater functionality than older versions. The example that most computing users see every day is that of "Windows," the Microsoft operating system.

Windows comes in Windows 3.1, Windows 95, Windows 98, Windows 2000, and Windows NT. Since its release, SAP has seen a number of versions of its R/3 software, including 3.0, 3.1, 4.0, 4.5, and 4.6.

Enterprise resource planning software comes in different versions because new features are incorporated as part of the software's evolution. There are both advantages and disadvantages of new versions for consumers of ERP software. The advantages include accessibility of new features and the elimination of previous bugs. The disadvantages include potential conflicts between different versions of the software and the costs of upgrading.

Software Choice and Requirements Analysis

This section reviews some basic issues in software choice and requirements analysis.

Software Choice: Cost–Benefit Analysis

Ideally, software choice is not random but rather the result of careful consideration of all the factors involved in the decision. Further, rational choice of one software package over another would typically include some sort of cost–benefit or "return on investment" analysis. Unfortunately, benefits may be fuzzy (difficult to establish and predict) and some costs may be hidden. In fact, costs and benefits may not be fully measurable until the system has had a year or so to gather information. As a result, many analyses of package software are frustrated by a lack of reliable data.

"As Is" versus "To Be"

There are at least two different models of an organization's processes: the "as is" model is used to model current processes whereas the "to be" model is used to model an organization's future, preferred processes. As part of choosing software, organizations develop "as is" models or "to be" models (or both) of their processes. Those models then provide a basis of choosing between different software.

Reengineering

Reengineering has led to two different philosophies: best practices and obliteration. This section distinguishes between the two.

Best Practices versus Obliteration

Reengineering, as first discussed by Hammer (1990), was designed to obliterate existing processes and then redesign processes that met the firm's needs in light of advanced technologies. Unfortunately, firms found that it was hard to obliterate existing processes and workers. Further, when they did obliterate they did not know what should replace the existing processes or how to develop those new processes.

Because it exploited existing processes that were known to work, "best practices" rapidly found acceptance as the way to choose a new process. Best practices are those considered to be the better or best ways of performing a particular process.

One source of best practices is an ERP system. Generally, best practices are captured in ERP systems as choices that must be made in the implementation. Enterprise resource planning systems generally have a number of best practices available, so that the software can be customized to "fit" each company that installs the software. For example, SAP's R/3 has over 1,100 best practices available. Owing to the availability of such a large number of best practices, virtually each implementation is different as the portfolio of best practices chosen varies from implementation to implementation.

A number of consulting firms have their own databases of best practices that are used to provide best practice alternatives in the support of implementing ERP systems. For example, Price Waterhouse (1995) implemented "Knowledge View" as a knowledge base of best practices. Such best practices databases are part of what is increasingly being called "knowledge management."

Summary

This chapter has provided some background information in the areas of computing (client server vs. mainframe), networks (capacity, standards, and security), databases (flat vs. relational files), software (operating systems, legacy software, package software, database management systems, and versions of software), software choice (cost–benefit analysis and "as is" vs. "to be"), and reengineering (best practices vs. obliteration).

References

ASAP [World Consultancy] (1996). *Using SAP R/3*. Indianapolis, IN: Que.

Frye, C. (1994). "With Financial Apps, DBMS Support Often Drives the Sale." *Software Magazine,* June, pp. 55–7.

Hammer, M. (1990). "Reengineering Work: Don't Automate, Obliterate." *Harvard Business Review,* July/August, pp. 104–12.

Juergens, M. (1999). "Technical Infrastructure." Unpublished presentation, University of Southern California.

Langdoc, S. (1998). "ERP Reality Check for Scared CIOs." *PC Week,* September 21, p. 88.

Markoff, J. (1998). "Oracle Database Takes Aim at Windows." *New York Times,* November 16, p. C5.

Price Waterhouse (1995). *Welcome to Knowledge View.* Dallas, TX: Price Waterhouse.

Radding, A. (1995). "Building a Better Data Warehouse." *InfoWorld,* November 20, pp. 57–62.

PART TWO

ERP SYSTEMS

3

ERP Systems Background

The purpose of this chapter is to provide some basic background information about ERP systems. As a result, we will address a number of specific questions.

- What is an ERP system?
- Who are the ERP vendors?
- What are ERP "partners"?
- What are some sample ERP modules?
- What does it mean to talk of "best of breed"?
- What are "add-ons" to ERP?
- What are ERP MAPs?
- How do ERP systems work?

Since SAP is the dominant ERP system, I will use it to illustrate some of the general ERP concepts.

What Is an ERP System?

In this book, ERP systems are computer-based systems designed to process an organization's transactions and facilitate integrated and real-time planning, production, and customer response. In particular, ERP systems will be assumed to have the following characteristics.

- ERP systems are packaged software designed for a client server environment, whether traditional or web-based.
- ERP systems integrate the majority of a business's processes.
- ERP systems process a large majority of an organization's transactions.
- ERP systems use an enterprise-wide database that typically stores each piece of data once.
- ERP systems allow access to the data in real time.

- In some cases, ERP allows an integration of transaction processing and planning activities (e.g., production planning).

Moreover, ERP systems increasingly are assumed to have the following additional characteristics:

- support for multiple currencies and languages (critical for multinational companies);
- support for specific industries (e.g., SAP supports a wide range of industries, including oil and gas, health care, chemicals, and banking);
- ability to customize without programming (e.g., switch setting).

Who Are the ERP Vendors?

Depending on who you talk to, the primary ERP vendors are referred to as BOPSE (**B**AAN, **O**racle, **P**eopleSoft, **S**AP, and J.D. **E**dwards). Other ERP firms include (but are not limited to) Great Plains, Lawson, Platinum, QAD, and Ross and Solomon (see e.g. Keeling 1996; Kersnar and May 1999). For additional information, see PricewaterhouseCoopers (1998).

BAAN ⟨www.baan.com⟩

BAAN was founded in the Netherlands in 1978. BAAN's ERP market share is roughly 5% (Stein 1997), and 1998 revenues were roughly $750 million (Bylinsky 1999). BAAN has approximately 3,000 clients in 5,000 sites worldwide. BAAN was thrust into the national ERP software spotlight when they won the Boeing ERP engagement in 1994. The founders recently left BAAN, in part because of irregularities in financial reporting that led to inflated sales figures (Maremont and Rose 1998).

Oracle ⟨www.oracle.com⟩

Oracle is the second-largest supplier of software in the world. However, they are perhaps best known for their database system, not their ERP applications. Oracle was founded in 1977 in the United States. Oracle's applications were developed for the U.S. market in 1989 and for the international market in 1993. In 1997, Oracle announced that they were going to market to specific industries (Greenberg 1997a) and improve the international characteristics of their software (Greenberg 1997b). In 1997, Oracle's market share was reportedly 10% of the ERP market (see Herrera 1999), and 1998 ERP revenues were $2.4 billion (Bylinsky 1999). Oracle reportedly can accommodate over 1,000 users (Keeling 1996).

Oracle has been criticized for being a database company and not an applications company. However, as noted by Kersnar and May (1999, p. 44), "Oracle's prowess in the database business makes its offering particularly attractive to firms that rely heavily on their own databases for competitive advantage." Oracle's reputation in ERP systems is for developing a product that can be interfaced with other products in order to construct a "best of breed" system. Oracle is likely to build software in-house (Holt 1998). In addition, possibly because of their basic focus on database management systems, Oracle was the first to provide a data warehouse product and the first to begin to integrate the Internet into their products.

PeopleSoft ⟨www.peoplesoft.com⟩

PeopleSoft was founded in 1987 and went public in 1992. PeopleSoft is the third-largest ERP vendor. In 1997 their share of the ERP market was 6%, and in 1998 their revenues exceeded $1.3 billion (Bylinsky 1999). PeopleSoft can be scaled to accommodate from 10 to 500 users (Keeling 1996).

PeopleSoft has become known for the broadest human resources capability (Kersnar and May 1999). In many cases, firms have chosen some other ERP (e.g., SAP) for all other modules and PeopleSoft for human resources. In some cases, the quality of this human resource module led some clients to adopt the rest of PeopleSoft's ERP modules.

SAP ⟨www.sap.com⟩

The largest market share for ERP is held by SAP (Systems, Applications, and Products in Data Processing), with estimates ranging from 30% to 60% of the ERP market share. In 1997 SAP had a 33% market share (Stein 1997) and supplied 60% of ERP used by multinational companies (Bowley 1998). In 1998, that market share had increased to 36% (Herrera 1999). SAP is the fourth-largest supplier of software, trailing only Microsoft, Oracle, and Computer Associates International. SAP was founded in 1972 in Walldorf, Germany. Never before has a company outside of the United States had such success (Edmundson, Baker, and Cortese 1997).

SAP is known for spending a large portion (typically 20% to 25%) of its revenues on research and development. Reportedly, SAP has over 9,000 implementations of R/3 at over 6,000 companies and over 2,500,000 users. In 1997, SAP's revenues exceeded $5 billion (Bylinsky 1999). Keeling (1996) reported that R/3 can be scaled for between 25 and 1,000 users.

SAP has a reputation for acquiring firms with features that they are interested in and then completely reprogramming those systems for integration with

R/3 (see Holt 1998). SAP is generally either the first or second of the BOPSE to adopt new features or capabilities. For example, they were the first to pursue specific industry versions of their software and the first to be truly international. In addition, they were among the first to pursue developments such as data warehouses.

J.D. Edwards ⟨www.jdedwards.com⟩

J.D. Edwards recently introduced its multiplatform software, OneWorld, which was designed to gradually replace its previous AS/400 product (see Keeling 1996). Historically, J.D. Edwards has been the leading supplier of AS/400 applications. OneWorld is now available on Windows NT, UNIX, and AS/400. OneWorld was rewritten in 32-bit technology for Windows NT and Windows 95. Reportedly, OneWorld is designed for between 5 and 500 users. In 1997, J.D. Edwards commanded about 7% of the ERP market (Stein 1997). In 1998, ERP revenues were $979 million.

What Are ERP "Partners"?

Enterprise resource planning firms do not implement all the software they sell. Instead they typically work with a wide range of partners in order to implement the software (although this also leads to interesting problems, as described in March and Garvin 1996). For example, SAP has four types of partners: alliance (professional services firms), platform (provide hardware), technology (provide operating systems and database systems), and complementary (tools that run with R/3). The approach used by SAP (and soon copied by other ERP vendors) was to leave a major portion of the implementation dollars on the table to be shared with their partners. In Europe, their plan was to share 80% of the revenues with the implementation partners; in North America, their plan was to share 90% of the revenues in order to generate market share (March and Garvin 1996).

The alliance partners are an interesting case because they indicate a rapid growth in the number of consultants. Kay (1996) reported that Andersen Consulting was the largest employer of consultants, with 3,200; SAP had 2,800, Price Waterhouse 1,800, and Deloitte & Touche 1,400. By 1998, the number of SAP consultants had grown substantially. For example, as noted in *Public Accounting Report* (1998), Deloitte & Touche had over 4,000 SAP consultants, almost triple their number of SAP consultants in 1996. Finally, by mid-1999, *Forbes* magazine (Herrera 1999) reported that there were roughly 50,000 consultants working on SAP engagements, with 10% working for SAP.

SAP's alliance partners now include Andersen Consulting, Cap Gemini, CSC, Deloitte & Touche Consulting, EDS, Ernst & Young, Hewlett-Packard, IBM, KPMG, PricewaterhouseCoopers, Siemens Nixdorf, and others. *Public Accounting Report* (1998) indicated that "services involving application software packages, such as SAP, Oracle, PeopleSoft, BAAN and Lawson, generate one third to one half of the total consulting revenue" at the Big 5 professional services firms.

SAP consultants generally were viewed differently than the alliance consultants at the consulting firms. For example, as noted in March and Garvin (1996, p. 7):

> There is nothing that we do that our partners do not. But you could best describe our consultants as deep and theirs as broad. Our consultants bring in-depth product knowledge, and are always a little further ahead than partners on the capabilities and requirements.

SAP is not the only ERP vendor with extensive ties to consulting firms. For example, as noted by *Public Accounting Report* (1997), each of the then Big 6 (now Big 5) professional services firms had consultants specializing in BAAN and other packages.

What Are Some Sample ERP Modules?

Enterprise resource planning systems can include a wide range of functionality, using components that are often referred to as "modules." However, there is some variance in different packages as to which modules are included and how they are named.

SAP's Modules

SAP's R/3 includes the following application-based modules (see e.g. ASAP 1996, pp. 74–8).

- AM (fixed asset management), which captures information relating to depreciation, insurance, property values, and so on.
- CO (controlling), which includes CCA (cost center accounting), PC (product cost controlling), and ABC (activity-based costing).
- FI (financial accounting), which includes GL (general ledger), AR (accounts receivable), AP (accounts payable), and LC (legal consolidations).
- HR (human resources), which includes PA (personnel administration) and PD (planning and development).

- MM (materials management), which includes IM (inventory management), IV (invoice verification), and WM (warehouse management).
- PM (plant maintenance), which includes EQM (equipment and technical objects), PRM (preventive maintenance), SMA (service management), and WOC (maintenance order management).
- PP (production planning), which includes SOP (sales and operations planning), MRP (materials requirements planning), and CRP (capacity requirements planning).
- PS (project system), which includes project tracking and budget management.
- QM (quality management), which includes CA (quality certificates), IM (inspection processing), PT (planning tools) and QN (quality notifications).
- SD (sales and distribution) system.

In addition, there are some cross-application (CA) modules that can be used throughout the R/3 system; these include SAP business workflow and SAP office.

Oracle Applications

Oracle's core applications fall into three primary groups: demand, supply, and finance. Demand includes order entry, accounts receivable, and inventory. Supply includes engineering, bill of materials, materials requirements planning, work-in-process, and purchasing. Finance includes general ledger, accounts payable, and cost management.

There are additional, incremental applications. Demand can also include sales commissions and sales compensation. Supply can include supply chain planner, supplier scheduling, and capacity and quality. Finance can include fixed assets, project accounting, and financial analyzer. Finally, other applications include human resources, payroll, data warehouse, and ad hoc reporting.

What Does It Mean to Talk of Best of Breed?

Generally, firms choose a single ERP package for implementation. However, a number of firms choose a "best of breed" approach where they attempt to mix and match modules in a way that best meets their needs.

Interchangeability

Different ERP system modules are not interchangeable. Moreover, there are few incentives for most of the BOPSE firms to make their software interchangeable with other, non-BOPSE firms. In addition, for the most dominant of the BOPSE firms there are even fewer incentives to make their software interchangeable with less dominant firms. As a result, we are not likely to see the days of inter-changeable modules in ERP systems in the near future.

Best of Breed

Although modules are not directly interchangeable, there are a number of situa-tions where firms might explore best of breed as an alternative. First, there may be a dominant ERP solution in the industry that seems to be missing one key set of features that could be accommodated by an alternative piece of software. Second, there may be multiple divisions with many similar but also some dif-ferent requirements. In this setting, a basic ERP system could be adopted while individual (divisional) needs could also be met through additional software. Third, perhaps no single ERP system meets the firm's needs, yet by mixing and matching software an appropriate solution could be found.

Advantages and Disadvantages of Best of Breed

The best-of-breed approach has one primary benefit: ideally the firm will get the system and functionality that they want with the heterogeneous system. However, there are likely to be a number of additional costs, including the fol-lowing. First, the search costs generally would be larger in a best-of-breed approach, since any module that might possibly meet the firm's needs should be considered. Second, the "look and feel" of the modules may differ; this may result in increased education or the need for reprogramming or repackaging. Third, the different modules will need to be interfaced with each other. This cost could be substantial. Fourth, best of breed is likely to require a diversified team, where the implementors' expectations are not likely to be as well spec-ified. Fifth, ERP systems have new versions every two to five years. With a best-of-breed system there could be timing issues as to when different modules have new revisions, which inevitably will be out of sync. Sixth, if different branches of the firm employ different solutions to the best of breed then there may be no commonality between systems, curtailing one of the primary advan-tages of ERP. Unfortunately, as noted by Freeman (1997, p. 61), in many cases "the costs of interfacing specialized manufacturing software in an integrated ERP system far outweigh the benefits."

What Are "Add-Ons" to ERP?

For each of the ERP systems, various vendors have developed a wide range of add-ons – additional software that provides increased capabilities. For example,

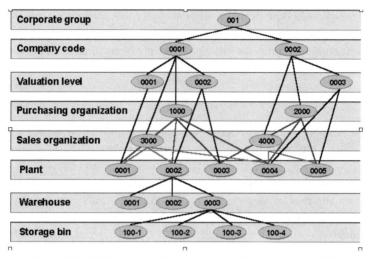

Figure 3.1. R/3's Organizational Structures Model (Source: SAP)

SAP has encouraged various third-party developers to build software to interface with SAP. Such software increases its functionality and provides insights into what new and emerging capabilities are of interest. A number of technologies (beyond the scope of this book) make these links possible, including application link enabling (ALE), object linking and embedding (OLE), and remote function calls (RFCs). In some cases, certification programs are available. For example, SAP certifies what they call "complementary" solutions, a number of which are reviewed in ASAP (1996).

ERP Models, Artifacts, and Processes (MAPs)

Configuring an ERP system means making choices about the models, artifacts, and processes (MAPs) that are embedded in the system and used by the organization.

Models

A number of models are embedded within ERP systems, such as the organizational structures model in SAP's R/3 (Figure 3.1). These models are representations of the world encompassed by the system, and their quality is important for capturing reality. For example, in the organizational structures model, the model allows capturing information down to the "bin" level. Similarly, information can be consolidated all the way up to the corporate group level.

There are both benefits and costs to such detailed capabilities. On the one hand, they can provide the system with the degree of organizational detail required to model the firm. On the other hand, if the model changes then it must be changed in the system. As a result, if the models are volatile then keeping them up-to-date can be quite costly.

There are specific assumptions made about each of those underlying models that must be accommodated as part of the implementation. For example, as discussed in the appendix to Chapter 9, R/3's organization model required Microsoft to make each department either a cost center or a profit center for valuation modeling purposes. Unfortunately, this view was different than the existing organizational model and so required that Microsoft adjust to the model.

Artifacts

Simon (1985, p. 10) defined an artifact as "an interface ... between an 'inner' environment, the substance and organization of the artifact itself, and an 'outer' environment, the surroundings in which it operates." The inner environment is the computer program and the outer environment is the world in which the system functions.

Enterprise artifacts are the food for information processes. An example of an enterprise artifact typically thought of as a "document" is an invoice. In addition, instantiations of models (in the form of charts of accounts, vendor lists, product lists, etc.) are also artifacts. Enterprise artifacts, commonly known as documents, are generated by systems as outputs (e.g. invoices) or used by the systems as inputs (customer orders). Enterprise artifacts that are instantiations of the models (e.g. vendor lists) provide structure to the enterprise systems.

Processes

Processes are the activity and information flows necessary to accomplish a particular task or set of tasks. Organizations must choose processes that will meet their needs, typically from among the portfolio of processes available within the ERP system. Generally, there are multiple ways to accomplish a task or set of tasks, so processes are not unique. Since they are not unique, it is expected that some processes will work better than others. Within ERP systems there are multiple cross-functional processes.

An illustration of SAP's order management process is given in Figure 3.2 (McAfee and Upton 1997). The process maps into multiple SAP modules that are integrated with each other. In contrast, a functional-based legacy system typically would have had at least four different systems (sales and distribution,

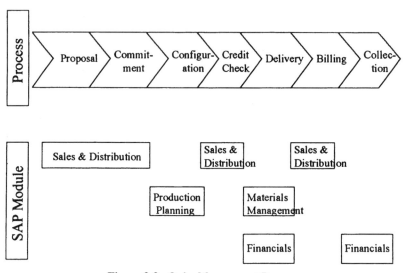

Figure 3.2. Order Management Process

production planning, materials management, and financial) that would not have been integrated. Instead, information probably would have been exchanged manually, if at all.

Enterprise resource planning implementation of processes ultimately requires many decisions, which are typically made by the implementation team. These decisions include, for example, who should get credit and who should be cut off, or when should discounts be offered and who should get rebates. Koch (1999) quotes an ERP implementation team member as follows: "We basically run the company. We make the decisions. Two years ago it would have been unheard of for [people in our positions] to make those decisions, but now people want us to run their departments."

How Do ERP Systems Work?

We will consider the case of SAP in order to gain a basic understanding of ERPs in general. An excellent example is given in Edmondson et al. (1997) and summarized in this section. International Sneaker Company (ISC) is a hypothetical U.S. company with worldwide sales; they manufacture their product in Taiwan.

(1) *Ordering.* A sales representative from ISC takes an order from a retailer in Brazil. Entering the data on her personal computer, the sales representative accesses R/3's sales module. The system checks the price as well as the discounts that the retailer is eligible for. The system also

checks the retailer's credit history to make sure that the firm wants to make the sale.

(2) *Availability.* R/3 software next checks the inventory. It finds that half the order is available from a warehouse in Brazil and so that portion of the order can be filled immediately. R/3 finds that the other half of the order will need to be delivered from ISC's factory.

(3) *Production.* R/3 alerts the warehouse to ship the portion of the order that is in stock to the retailer. In addition, R/3's manufacturing software schedules the production of the remainder of the order. An invoice is printed up in Portuguese.

(4) *Manpower.* When scheduling production, R/3 notes that there is a shortage of workers to handle the order. It alerts the personnel manager of the requirement to hire temporary workers.

(5) *Purchasing.* R/3's materials planning module notifies the purchasing manager that it is time to order new raw materials and also of the amounts that need to be ordered.

(6) *Order Tracking and More Ordering.* The Brazilian retailer logs onto to ISC's R/3 system through the Internet and sees that a portion of the order has been completed. In addition, the retailer uses this as an opportunity to place yet another order.

Summary

This chapter has provided some basic discussion of ERP vendors and systems to facilitate understanding of the subsequent chapters and to provide basic understanding of the ERP industry. The partnership arrangements that take place between ERP vendors and consultants and other support partners have been traced. Enterprise resource planning modules were investigated, with SAP and Oracle used as examples, and ERP MAPs were defined. Finally, a sample use of an ERP system (SAP) was traced.

References

ASAP [World Consultancy] (1996). *Using SAP R/3.* Indianapolis, IN: Que.

Bowley, G. (1998). "Silicon Valley's Transplanted Sapling." *Financial Times,* March 27.

Bylinsky, G. (1999). "The Challengers Move In on ERP." *Fortune,* November 22.

Edmondson, G., Baker, S., and Cortese, A. (1997). "Silicon Valley on the Rhine." *Business Week,* November 3, pp. 162–6.

Freeman, E. (1997). "ERP Recipe?" *Datamation,* August, pp. 61–4.

Greenberg, I. (1997a). "Oracle, PeopleSoft Target C/S Apps Vertically." *InfoWorld,* January 13, p. 10.

Greenberg, I. (1997b). "Oracle's App Upgrades Aim at Multinationals." *InfoWorld,* April 14, p. 14.

Herrera, S. (1999). "Paradise Lost." *Forbes Global,* February 8, ⟨forbes.com⟩.

Holt, S. (1998). "PeopleSoft Hops on Front Office Band Wagon." *InfoWorld,* August 31, p. 16.

Kay, E. (1996). "Desperately Seeking SAP Support." *Datamation,* February, pp. 42–5.

Keeling, D. (1996). "A Buyer's Guide to High End Accounting Systems." *Journal of Accountancy,* December, pp. 43–52.

Kersnar, K., and May, M. (1999). "Plug and Pray." *CFO Europe,* June, pp. 40–50.

Koch, C. (1999). "The Most Important Team in History." *CIO Magazine,* October 15.

March, A., and Garvin, D. (1996). "SAP America." Report no. 9-397-057, Harvard Business School, Cambridge, MA.

Maremont, M., and Rose, M. (1998). "Dutch Software Firm Has Extensive Links to Founder's Interests." *Wall Street Journal,* July 7, p. A1.

McAfee, A., and Upton, D. (1997). "Vandelay Industries." Report no. 9-697-037, Harvard Business School, Cambridge, MA.

PricewaterhouseCoopers (1998). *Technology Forecast, 1999.* Menlo Park, CA: PricewaterhouseCoopers.

Public Accounting Report (1997). "D&T Allocates 200 Consultants to BAAN Launch." February 28, p. 3.

Public Accounting Report (1998). "Big Six Dominate Systems Integration Market." July 31, p. 4.

Simon, H. (1985). *The Sciences of the Artificial.* Cambridge, MA: MIT Press.

Stein, T. (1997). "SAP's Fast Track." *Information Week,* September 1, pp. 14–16.

Appendix 3-1

Geneva Steel: Changing the Way Business Is Done

Joseph Cannon, chairman of Geneva Steel, felt that his company needed a change in culture in order to drive a change in the way it did business.[1]

> We made the decision to go with SAP. We might have made that decision anyway, but part of that decision and maybe a substantial part of that decision was because one of the ways you can drive culture change is to drive the information flows, the availability of information, who can know what, how much you can know about how many different things ....
>
> So we thought, How do we move a ... traditional steel culture to more of a mini-mill type culture that would be flatter with greater information flow, less fear, more teamwork?
>
> How do you do that? Of course ... nothing is a real silver bullet. One of the drivers in any case was that we needed to change our systems .... We have a really outdated and primitive accounting system; that needed to be changed.

---

[1] Numerous quotes from Joseph Cannon were taken from an interview recorded October 1998 with Eric Denna and presented at the Technology Visioning Conference (November 19, 1998).

SAP in the Steel Industry

By 1999, SAP's enterprise resource planning system (R/3) had been successfully adopted by a number of firms in the steel industry. Of particular note to Geneva Steel was that, in 1995–96, the Swiss firm Kindlimann AG implemented SAP's R/3 (release 2.2) using the modules of financial accounting and controlling (FI/CO), logistics with sales and distribution and material management (SD/MM), and human resources and asset management (HR/AM). At that time, a member of the management board at Kindlimann noted[2] that the

> SAP implementation was a strategic move for us to retain our leading market position in the future. The competition is increasingly opting for information technology, which directly affects price, quality and customer contact. SAP is the means to optimize and coordinate these factors, combined with the enthusiasm and commitment of all our employees.

New Steel (May 1997) reported that LTV, Hylsa, Bohler, Arbed, Sidmar, Preussag, Thyssen, Krupp-Hoesch, Mannesmann, Hoogovens, Kindlimann, and BHP had all purchased SAP's R/3 software. In addition, Acme Steel was reported in 1998 as implementing SAP in order to handle their invoices, billing, and scheduling and tracking orders, including using the Internet for taking orders.

Geneva Steel

Background and Short History[3]

Geneva Steel, located in Vineyard, Utah, is the only integrated steel mill west of the Mississippi River; in 1998, Geneva had over 2,400 employees. Geneva Steel's customer base was primarily in the western and central United States. Geneva sells its products to steel service centers and distributors, such as Mannesmann Pipe and Steel. Roughly three fourths of Geneva's sales are sheet and plate steel; the remainder is pipe and coil.

Cut-to-length plate is used in the manufacture of ships, barges, and industrial equipment. In addition, it is used in construction of a wide range of products that include roads, bridges, and buildings.

During World War II, the U.S. government's largest construction project was the development of the Geneva Steel plant. The plant's inland location was chosen because of the proximity to the necessary raw materials and as a precaution against a Pacific coast invasion or closure of the Panama Canal. In

---

[2] Kindlimann AG, by SAP (no date).
[3] Based on ⟨eddy.media.utah.edu⟩ and ⟨www.geneva.com⟩ and Geneva Steel discussions in "Geneva Steel History," ⟨www.geneva.com⟩.

Table 3.1. *Geneva Steel Company Statements of Operations*

	1998	1997	1996
Net sales	720,453	726,669	712,657
Cost of sales	659,132	665,977	662,058
Gross margin	61,321	60,692	50,350
Selling, general, and administrative	22,116	22,488	24,943
Write-down of impaired assets	17,811	—	—
Income (loss) from operations	21,394	38,304	25,729
Other income (expense)			
Interest and other income	356	413	
Interest expense	(42,483)	(40,657)	
Income before provision (benefit)			
for income taxes	(20,733)	(2,040)	
Provision (benefit) for income taxes	(1,790)	(772)	
Net income (loss)	(18,943)	(11,608)	(16,327)

Note: Figures are given in thousands of dollars.
Source: Company press release of November 16, 1998 and 12/97 10-K.

December 1944, the plant was opened and operated for two years as a U.S. government facility. Geneva's steel was used to help construct more than 2,000 Liberty class ships. These ships were designed to last about five years and were used to deliver food, clothes, and ammunition overseas.

After the war, the federal government solicited bids for the plant. U.S. Steel won and ran the plant for over forty years. However, early in 1987, the plant was closed owing to a variety of factors, including increased foreign competition and higher labor costs.

In 1987, Chris and Joseph Cannon managed to purchase and reopen Geneva Steel under the new ownership of Basic Manufacturing and Technology of Utah. As of January 1999, Geneva's chairman (Joseph Cannon) and president (Robert Grow) owned over half of the company.[4] Table 3.1 summarizes operations for the years 1996–98.

Geneva had a traditional steel mill–like culture. As noted by Cannon, "[w]e had a culture that was a lot like traditional steel mill culture. That would be hierarchical, with quite a bit of fear. That is, fear of failure, fear of challenging people that were your superiors in the organization." Jacobson & Associates, who conduct a survey of steel customer satisfaction, recently found Geneva Steel had some problems in that area. As noted by Cannon, "[w]e were dead last."

---

[4] Hoover's Company Capsule.

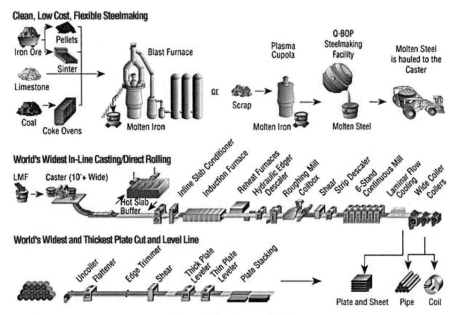

Figure 3.3. Geneva Steel

Production Process[5]

Geneva Steel uses iron ore, limestone, and coal to generate molten steel (see Figure 3.3). Using the world's widest in-line casting and direct rolling process, Geneva can produce steel slabs 126 inches wide. Specific finishing stands allow Geneva to roll the steel according to customer specifications, ranging in thickness from 1 inch to 0.07 inch. Geneva also has the world's widest and thickest plate cutting and leveling line, which is used to facilitate generation of those products. Geneva's production process is summarized in Figure 3.3.

Mini-Mills[6]

During the period 1970–85, the U.S. steel industry reduced its capacity by about one third. While several large companies were reducing their capacity, a group of small steel companies doubled their output using EAFs (electric arc furnaces) or "mini-mills." By 1991, mini-mills accounted for 22.7% of U.S. steel production, up from 15% in 1981. Some experts believe that mini-mills will ship 14 million tons of hot rolled steel in 2000, up from 3 million tons in 1995.

---

[5] Based on Geneva Steel's discussion ⟨www.geneva.com⟩.
[6] Based on ⟨www.steelnet.org⟩.

As noted by SteelNet,

> EAF industry growth can be attributed to several factors, the most significant of which are cost efficiency and quality. With low operating costs, and efficient, flexible organizational practices, EAF steel producers are able to expand cost advantages and product quality.

Geneva Steel's Management Information Systems

Geneva Steel's information system configuration evolved over time to include a number of heterogeneous systems that made integration and information access difficult. As noted by Cannon,

> we have ... a mainframe ... [and] ... a primitive accounting system ....
>
> [W]e have lots and lots and lots of different kinds of computers. They have a hard time talking to each other. We have a large number of mini computers out there that are different kinds, that have different software .... Our system is a road map from hell.

The current information system generates substantial amounts of transaction data, but the data from most transaction systems are not generally available. As noted by Cannon, "[We] ... generate literally millions of datapoints. But ... they are not very accessible to us, and they certainly are not integrated with each other."

The information system provides three primary daily reports to support Cannon's decision making.

> One of the primary reports which I look at,... is the Operations Report. It will list ... yields,... pretty specific things ... [for example] ... number of bars per hour. It is at a lower level of operational detail than the DOR (daily operations report) or the DRO (daily report of operations) .... And those are only three of the reports that I get. I'm not talking about sales reports ... that are separate .... We also generate thicker reports on lots of things other than sales, to accounting, to shipping, and to operations. But the reports that I just mentioned ... are daily reports. Then we have some that are at least weekly, and maybe a little more often ... for sales information. The good thing is that I get most of this on e-mail and so I don't actually have lots of paper. But it is true, there are separate reports.

Unfortunately, the information system provides limited knowledge of customers and their orders.

> There may be things we don't know about the costs of hot metal coming in that are that way because of a particular customer. I think that we have an intuitive sense of that, and maybe in some cases we have an actual sense of that.
>
> Inside sales representatives ... have a hard time figuring out where an order is.

Furthermore, the current information system limits the links with customers. As noted by Cannon,

> How do [customers] ... get information? How do they know when a piece of steel is coming in their direction? Is it with the order? Is it going to get there when they ordered it? Those are all things that can be improved by greater communications with them and greater communications internally.

In the current information system, information integration is a critical part of management's job.

> Integration is mostly mental. In other words,... managers are the integrators. I look at sales, and I look at DORs, and DROs. Mostly I should say in fairness we use the DROs primarily, it is just that we generate a lot of other reports ....
> But on a daily operating basis, the integration is in the mind of the person and hopefully that person working with other people.
>
> Further, [when] ... we do a business plan ... we have to struggle with integrating information ....

Currently, different legacy systems have personnel effectively responsible for "interface gates" – the interface between care and feeding of a program and its users and other programs. Few legacy programs talk to each other. Moreover, only a limited number of technology people are familiar with particular applications, and only a few accountants understand what the programs do and what information is required for their use. As noted by Cannon, in essence "[y]ou ... need to have a cost accountant for every single department."

Geneva Steel's Delta Project

In 1997, Geneva began a major reengineering effort with their "Delta Project." The purpose of the Delta Project was to "effect systemic and pervasive change with respect to corporate systems, processes and structures." The goal was to reduce costs by 20% in comparison to calendar year 1996. As noted by Cannon in a March 1997 press release,

> [o]ur efforts to reduce administrative costs are only one element of a larger project referred to as the Delta Project, directed at systematically changing the way we do business at Geneva Steel. During the past several months, we have launched several teams that are focusing on improving all aspects of the company, including customer service, employee relations, and operations. These initiatives are necessary to remain competitive in the current steel market place. Recent surges in steel imports, combined with increasing domestic capacity, have intensified competition in our markets. We are responding by making systemic, pervasive changes in order to assure our long-term competitiveness.

The Delta Project began to generate a number of changes.

Organizational Changes

On June 2, 1997, Geneva Steel announced a realignment of management responsibilities among senior executives. According to the company press release,

> The changes in our management structure are intended to focus our resources
> on those areas with the potential of yielding the greatest improvements to the
> company .... The changes in reporting relationships are intended to reflect the
> Company's commitment to a flat streamline organization focused upon imple-
> mentation of our strategic plan.

Since September 1997, Geneva's administrative and executive staff has been reduced from 328 to 219; operations management was reduced from 155 to 92.[7]

Organizational Measures

In order to assess the impact of the Delta Project changes, Geneva developed
a number of criteria to provide a balanced measurement of quality, timeliness,
and cost. Primary concern was given to rate of improvement in the following
metrics:[8]

- on-time deliveries;
- order fulfillment accuracy;
- manufacturing cost per ton;
- administrative cost per ton;
- customer complaints;
- manufacturing yields by activity;
- inventory levels and product.

As noted by Cannon,

> [a]nother metric would be customer satisfaction – how do our customers observe
> us. And we weren't doing this nearly as well as we could have or should be doing
> .... We are looking to custom integration as well as some organization changes
> to address that.

SAP

Unfortunately, Geneva's current management information system did not capture sufficient information to allow accurate measurement of each metric. In
addition, their legacy systems limited adoption of new workflows. As a result,

---

[7] Geneva Steel, December 1998 10-K.
[8] Eric Denna, "Emerging Economic Entities," presentation made at the AICPA/AAA Technology
Visioning Conference (November 19, 1998).

Geneva chose to implement SAP in order to facilitate new workflows and enable capture of information for the new balanced set of measures.

Geneva Steel's Expectations of SAP

As Joseph Cannon looked ahead to a complete implementation of SAP's R/3, there were a number of expectations.

Fewer Hardware and Software Systems

SAP adaptation is implemented in a client server environment. Hence, Geneva's mainframe would not be a part of the system implementation. In addition, implementation of SAP typically reduces legacy applications anywhere from 50% to 90%. And since SAP is an integrated application, there would be fewer hardware and software systems; the SAP system itself would be fully integrated.

Reporting Capabilities

Cannon anticipated that SAP would provide more robust and integrated reporting capabilities: "One of the virtues of SAP is that it puts all of [the sales, DORs, and DROs reports] … together, you don't need to have three different reports."

Information Availability

Cannon indicated that information access should improve with SAP:

> my expectation of SAP is that then you have an integrated tool … that by itself isn't self-executing, it won't do anything in itself, but we can have greater access to our information in a more understandable, actionable, and usable form.

Cannon also felt that SAP would make information (e.g., sales information) more readily available both internally and ultimately to customers: "If you have a part number and a system that … relates to itself, then people in different parts of the organization can know more information about what they need."

Uses of Information

Cannon hoped that SAP would generate more strategic use of information, facilitating the asking of "what if?" questions, and so improve decision making.

> Hopefully [our information usage will be] … more strategic …. I do think it puts a tool/weapon in [management's] … hands to get us alerted earlier:… "have you thought about this?"

> I think it gives us a greater opportunity to ask the questions "what if you try this?" or "what if you try that?" I think that we'll have a database there that we can say, "what happened when we did this?", and we can go back and look.

In addition, Cannon felt that there would be increased access to data and, as a result, increased ability to perform analyses of relationships between groups of data:

> There will be many, many more datapoints .... So this way you can relate two variables. You can relate the sale of a particular product maybe much more directly and accurately to the entire cost system.

Finally, Cannon felt that SAP would provide for increased capabilities – such as question asking – that would not be fully known or understood until the system was actually implemented: "this would just give you a larger platform playing field to know a lot of questions, some of which I don't know yet, but some of which will come up when we start looking at things."

System-Based Integration of Information

Rather than depending on managers to be information integrators, Cannon also felt that SAP would more tightly integrate across business functions.

> We will know what the cost is. Why will we know that? Because the purchasing people will have done their thing. The human resources people have done their thing ... and all of that will be integrated in a system that will be that much more transparent. You won't need to have a cost accountant for every single department.

Organizational Interfaces in Accounting and Information Systems

Cannon expected that implementation of SAP would lead to a tighter link between accounting and information systems, as well as an integration of the accounting function with the management team.

> [I expect our] accounting professionals, our CFO and controller are part of that team, [to work] with the management team as a whole .... I expect a really close integration between accounting and information systems. And that that would produce ... a framework for making decisions.

In addition, implementing SAP is expected to have some effects on the number of accountants and information systems personnel. As noted by Cannon,

> Well, in the first place we will have fewer accountants and probably have fewer information systems people. Because one of the things we are considering is contracting out a chunk of that function. A great deal of what ... we have cost accountants do ... won't need to be done, once SAP is implemented.

As a result, Geneva expected their staff to drop from roughly 80 to 12 in the information technology department and from 60 to 10 in their accounting function.

Planning and Information

Cannon also felt that the system should affect customer satisfaction and the company's ability to plan: "Another huge impact for the [SAP] system, [is] how we plan .... How you plan your production come-downs, the production process system is very important. That is [just] one aspect of it."

Geneva Steel's Introduction of SAP

Geneva Steel disclosed in their December 1997 10-K that they were beginning implementation of SAP's R/3. Geneva Steel's 1998 SEC forms 10-Q for February 13, May 15, and August 14 all contain the following statement.

> The Company has selected and started the implementation of SAP software, an enterprise-wide business system. The Company expects to benefit significantly from such implementation, including addressing the year 2000 issues inherent in its mainframe legacy systems. The project is currently estimated to cost $8.0 to $10.0 million with implementation completed in 1999. Review is underway to assess and subsequently address any other impact of year 2000 issues on process control programs and hardware. Depending on market, operation, liquidity and other factors, the Company may elect to adjust the design, timing and budgeted expenditures of its capital plan. In addition, the Revolving Credit Facility contains certain limitations on capital expenditures.

In their December 1998 10-K, Geneva noted that

> This system [SAP] affects nearly every aspect of the Company's operations. During fiscal year 1998, the Company installed new year 2000 compliant HP computer hardware and SAP modules for financial accounting, purchasing and accounts payable, raw materials, inventory control, and accounts receivable. The Company is performing final integration, testing, and validation on various other SAP modules. The human resource and payroll module is expected to be implemented on January 1, 1999. In early 1999, the Company will also implement other SAP modules, including sales and distribution, materials management, production planning, and product costing and other management information systems.

Geneva Steel also indicated that, as of September 30, 1998, they had spent approximately $5 million implementing SAP and that the implementation would be completed in 1999.

Acceptance of and Resistance to SAP

Acceptance of SAP was neither immediate nor universal. As noted by Joseph Cannon,

I've seen even the beginning of ... cultural change. Obviously there were people who greatly and deeply resisted moving to SAP. "Oh it costs a lot of money," "It's just a big hassle," "You are just buying a new toy," .... I have heard all of those and more in discussions about why people resist it .... I don't think, even today, that we have a total conversion .... I think we got a critical mass of people who are supportive. So when you look at the training already that's going on here, you are starting to see a good deal more enthusiasm in SAP. And, as a matter of fact, people have now finally figured out that whether you didn't like or whether they liked it, it is going to happen.

Questions

1. What problems does Geneva have with their legacy systems?
2. What expectations do they have for the SAP implementation?
3. Can SAP meet Geneva's expectations?
4. What do you think of the metrics that Geneva planned on using to evaluate the Delta Project?
5. Geneva plans to use SAP to change the culture at Geneva. What are the advantages and disadvantages of this approach to culture change?

4

ERP Data Input

Enterprise resource planning systems ultimately result in the reengineering of organizational processes. For example, in many legacy systems data is gathered at the loading dock, filtered through accountants, and then entered into the system. However, ERP systems are designed so that their usage can be pushed to the point of data generation, often in operations. As a result of this reengineering (see Hammer 1990), there are a number of changes in the process, including who gathers the data, how it is gathered (actually gathering more data, bypassing paper and entering data straight to a computer-based environment), gathering the data where it is generated, removing accountants and replacing them with information gatherers from operations, and changing when the data is generated to correspond to a process focus.

Each of these system changes can have a significant impact on capturing data inputs (e.g., who does data input, where is data input done, and how often is input data gathered), and this ultimately influences the quality of the data. As a result, reengineering changes to data inputs can have a major impact on the corresponding benefits and costs of an ERP implementation.

Assessing benefits is often more difficult than assessing costs – probably because benefits are not yet actualized and often are less direct. As a result, in order to understand the contribution of benefits, those associated with system changes need to be identified.

On the other hand, implementation costs can often be readily identified. Even so, there are some costs that are more difficult to measure. For example, changes in data inputs can engender user resistance. Since user resistance can influence the implementation cost and potential success of the ERP implementation, it is important (a) to identify emerging issues that can generate user resistance and (b) to develop a model that can facilitate understanding the effects of implementation difficulties deriving from data inputs. For example, since ERP systems are more general than legacy systems, the screen and data

Table 4.1. *Sources of Benefits*
and Costs Due to Data Inputs

Gather data once
Gather more data
Gather data straight into the computer
Gather data where it is generated
Gather data with a process focus

input structure can cause ERP data input to be slower than that using legacy systems, which were often designed specifically to accomplish a particular purpose. The time required to input data to multiple ERP screens can have an effect on the number of transactions that can be processed during any given span of time. As a result, in those organizations where there are substantial differences between legacy and new ERP capabilities, processes such as order processing may be materially slower because of data inputs. Slower processing of inputs can lead to decreased throughput, larger resource expenditures, and more difficult implementations. It is therefore important to have a model that can allow identification of such difficulties. Accordingly, this chapter presents two such models.

In any case, user resistance to implementation due to data input problems has led ERP vendors to devote increasing attention to making their systems easier to use. However, based on the way the ERP systems historically have been designed and built, a number of experts suggest that the problems with data input will continue.

Sources of Benefits and Costs Due to Data Inputs:
The Impact of Reengineering

Enterprise resource planning systems facilitate many of the principles of reengineering enumerated by Hammer (1990) and others; some of those principles ultimately affect data inputs. Although reengineering can improve organizational processes, the sources of benefit can drive user resistance. These sources of change are summarized in Table 4.1.

Gathering Data Once

There are a number of benefits for gathering data at a single time, as opposed to multiple times. Generally, cost is lower if the data is gathered a single time. If data is gathered a single time then all users are "on the same page" with data uses, so that forecasts and the like will be done using the same data. Updating data costs are lower, since only one instance of the data will need to be updated.

Enterprise resource planning systems typically employ a relational database structure, so that once a piece of data has been gathered it does not need to be gathered again. Gathering data only once will ensure that all appropriate uses and users have access to the same data value at the same time, ensuring that the same data is used throughout the firm for planning and control purposes. In addition, since data is gathered only once, firms see ERP systems as providing an opportunity to reduce the number of clerical workers and corresponding costs. Clerical workers and their co-workers are not likely to see this as an advantage, and this can be expected to increase user resistance.

The primary downside of gathering data once is that – if the data is incorrect – then all users of that data will be adversely affected. The ERP system does not have redundancy built into it. As a result, data input controls (to make sure that the data is correct) and organizational process maps (indicating job functions) are critical. Once incorrect data is in the system, it may never be discovered that the data is incorrect. For example, as noted by Koch (1999),

> pre ERP, warehouse clerks knew they could let a truck leave the loading dock without checking off the goods shipped on the packing slip; the slip would be there and if the clerks forgot about it, at some point, accounts receivable would call them up and yell. Not anymore. If the clerks don't account for everything when the truck leaves, the customer will never get an invoice, because the ERP system has no record of the goods being shipped. Accounts receivable won't ever know that the customer received the goods and won't be able to act as a sweeper upper anymore – no more wake-up calls to the loading dock.

Gathering More Data

Because all the data now reside in a single location, tied together in a relational database, it is now cost-effective to gather more data. As a result, rather than aggregating data prior to data input, it is now feasible to gather the data in a disaggregated manner and let the system do the necessary aggregation when reports are generated.

Unfortunately, as more data is gathered, the cost of gathering data increases. However, the decision-making benefits may outweigh any additional costs.

Gathering Data Straight into a Computer-Based Environment

Not only are the amounts of data and the number of times that the data is gathered changing, but the media of data collection is also changing. As noted by the SAP R/3 project manager at Purina Mills, "[w]e are taking folks who have recorded some information on pieces of paper and putting them on PCs [personal computers]" (Stedman 1998b).

The benefits of this approach can be high, for now there is no intermediate data replication (paper to paper to data entry). Instead, data goes straight into the system. However, there are also costs. For workers with limited experience with computers, such a change can be difficult and can influence the success of the implementation. In addition, this approach can in some cases make managers and others function as data entry clerks. Such a shift can lead to user resistance and a resultant negative impact on the implementation.

Gathering Data Where It Is Generated

Legacy systems generally employ a broad range of accountants and clerks who gather data for input into a system. Paper-based communications are sent to accountants, who ultimately are responsible for getting the data into the system. With an ERP, however, data gathering is pushed closer to its point of origination. As noted by the ERP project director at Purina Mills, "we have more people entering data than ever before" (Stedman 1998b). As a result, the personnel doing the data acquisition are not accountants but instead are operations people. In some companies this shift is expected to reduce the number of accountants and clerical personnel, as seen in the Geneva Steel case discussed in Chapter 3.

Competing forces determine whether this shifting of the data entry will improve quality. On the one hand, data can be more accurate because it is not required to go through multiple levels in an organization before it is input, as required in legacy systems. For example, in many legacy systems, data is gathered on paper at the loading dock, filtered through accountants, and then entered into the system. With an ERP system, it is not unusual to gather data at the loading dock that goes straight to general ledger accounts and the financial statements.

On the other hand, clerical accountants have sought out jobs that deal with data gathering and massaging information. In many cases, personnel (such as loading dock workers) will not be interested in or capable of dealing with data inputs for ERP. As a result, the personnel inputting the data may influence the quality of the data or may need additional training. Accordingly, the shift to gathering information at the source, from a broad base of workers, is not costless.

Thus, system changes may be required to mitigate user resistance. For example, as noted by the director of information management for Hydro Agri,

"SAP's user interface was confusing to loading dock workers, who enter the quantity of chemicals coming in or going out" (Stedman 1998b). As a result, Hydro Agri had to build an interface to improve usability and accessibility. Such problems can slow implementation and generate user resistance.

Gathering Data with a Process (Rather Than a Function) Focus

In legacy systems, data acquisition typically is based on particular functional application needs. However, in ERP event-based systems, data acquisition derives from a cross-functional, process focus. As a result, if the information is gathered at the source then the sequence of data origination is likely to change with a process focus.

Furthermore, with legacy systems the implicit model of the firm is likely to be more functionally based, with a focus on specific functional information needs. A move to a cross-functional set of events can yield a broader-based model of the firm, satisfying a greater range of informational needs. Enterprise resource planning systems like SAP model the firm at one level with a set of certain event "triggers" – these represent different events that determine when information needs to be gathered for the system. As discussed in Whang, Gilland, and Lee (1995, pp. 8, 26), SAP triggers may include:

- sales activity from contact of customer;
- goods receipts from purchase orders;
- reservations of material for planned use;
- goods issues (e.g., withdrawal of material);
- transfer postings (change in batch number, title change among departments);
- stock transfers (e.g., plant-to-plant internal move); and
- goods movement for production orders (parts out of the warehouse, finished goods into the warehouse).

In some cases, the event focus may change when information is gathered, who gathered it, and the sequence as compared to when the legacy system gathered that same information. This focus on trigger events has both benefits and costs. Benefits include the potential for a better model of how the company should function in the world. For example, a legacy system–based functional approach may focus on events of concern primarily at the functional level and not on broader information needs. A shift to a cross-functional process could change the event base of the system to better model the firm. Costs could include changes in the status quo that might lead to user resistance.

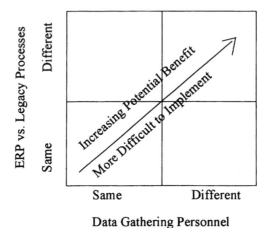

Data Gathering Personnel

Figure 4.1. Changes in Process and Data Gathering Personnel

Process Change and Data Personnel Change: User Resistance

Changes due to reengineering with the ERP system can lead to changes in process and personnel responsible for data entry. As processes change, there are concomitant changes in how data is gathered, how often data is gathered, where data is gathered, and who gathers data. If processes don't change much from legacy to ERP and if the assignment of people to data input doesn't change much, then the ERP implementation will likely have few difficulties with data input. However, if the process changes and the personnel collecting data changes then implementation becomes more problematic owing to issues such as user resistance. Unfortunately, it will be impossible to generate the benefits associated with an ERP system unless there *are* changes in processes associated with gathering data. Consequently, there is tension between the benefits of changing and the disruption that can occur in an organization as a result of accommodating the changes. This is summarized in Figure 4.1.

Too Many Screens and Too Much Time to Input the Data?

Enterprise resource planning systems are general packaged software designed for implementation across a broad base of firms. In order to provide system flexibility and capture a broader base of information, ERP system input screens are based on providing very general design capabilities. Not surprisingly, many firms do not need all of the data capture capabilities that these large ERP systems provide. However, configuration may not compress the number of screens that must be filled in. For example, with SAP's R/3, versions prior to 4.0 were less flexible in moving data fields to a single screen. As a result, data entry in an ERP can take more screens than if the screen sequence were specifically developed to meet the particular data entry needs. More screens means more data, more time, and thus more personnel needed for data inputs. Systems specifically designed to meet a particular company's needs may thus require less data input time and resources than an ERP system.

Hydro Agri

At Hydro Agri's Canadian stores, it used to take about 20 seconds to process a farmer's order. However, after they installed SAP's R/3, the processing time went to roughly 90 seconds (Stedman 1998b). Since Hydro Agri has about 30,000 orders every four weeks, the new system was requiring huge resources for data input. Prior to R/3, every four weeks required 600,000 seconds of order processing time; after R/3, 2,700,000 seconds. As a result, information technology staff were forced to take orders.

With R/3, workers had to provide data input into six screens; in the previous (home-grown) system, there was a single data entry screen. In addition, work processes as defined by R/3 were significantly different, so that even the workers responsible for data input were different than with the home-grown system. Consequently, the change to the new system forced additional educational requirements.

According to Hydro's director of information management, the increased requirements on order entry "threatened to be a showstopper." Thus, Hydro Agri was forced to choose between installing a third party's order entry package or developing their own custom package. Ultimately, Hydro Agri's approach to solving this problem was to build an application between R/3 and the users to simplify and consolidate access.

Why Did the Number of Input Screens and Processing Time Increase?

The change from a functional to a process focus changes the sequence of information gathering and who gathers the information, resulting in a different number of screens to input the data. For example, at Hydro Agri, the R/3 system had six screens whereas the legacy system had only one screen. The ERP system was built with a process focus in anticipation of the information deriving from multiple sources as part of this process focus – hence the need for multiple screens. However, the legacy system was built to exploit functional information specialization. As a result, in this case there may not have been a good match between the existing organizational process and the software. We

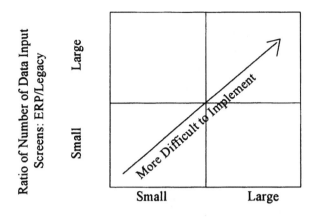

Number of Data Entry Transactions

Figure 4.2. Number of Input Screens and Transactions

will examine the relationship between organizational processes and software in more detail later in the book.

Number of Input Screens and Transactions:
A Model of Implementation Difficulty

If the number of screens in the new ERP system is substantially larger than in the legacy system, the result can be an increased amount of time required to input the data for a particular transaction. This can result in lost orders or an increase in resources necessary to support the data entry process. If the number of transactions is small then any increase due to increased data entry work generally will have a limited impact. However, if there are a large number of transactions, then a substantial increase in the number of screens (over the legacy system) will result in a substantial increase in resources required to get the data entered. As a result, if the ratio is high (e.g., at Hydro Agri it was 6 : 1) and the number of transactions is large (30,000 at Hydro Agri), then the system will be more difficult to implement unless the proliferation of screens is reduced (as was done at Hydro Agri). This increase in implementation difficulty is summarized in Figure 4.2.

The ratio of the number of data input screens in ERP systems to that in the legacy system may also provide a measure of "fit" between the ERP system and existing organizational processes. A large ratio is likely to indicate that the organization processes and the software processes are substantially different, possibly indicating a lack of fit of the software to existing processes.

ERP System Design

Enterprise resource planning input data differences when compared to legacy systems are driven by ERP system design issues, including the importance of the user interface in those designs, the focus on the database, and the relationship between processes and data inputs.

Insufficient Focus on User Interface

Some practitioners have argued that ERP developers have not spent sufficient time considering the user interface. As noted by the president of Human Factors International:

> The developers of these [ERP] packages, without exception, approach things from a system point of view, not a user point of view .... [The system forces the users to decrease productivity by requiring "window thrashing" – going back and forth between keyboard and the mouse.] ... The software drives people back and forth, and it's driving them nuts. (Stedman 1998b)

Other practitioners, such as an analyst from AMR Research, have argued that the problem of difficulty of ERP input results because ERP developers generally have "designed from the database out, not the user interface in. The screens have been the last part of the whole process."

Processes and Data Inputs

In some cases, loss of productivity in the data input arena is not as important as gains in the availability of decision-making information and changes in other processes. As noted by the director for communications at Procter & Gamble: "Better productivity in order entry and receipt of raw materials isn't a primary goal .... The real benefits are for business planners, material managers and other users, farther down the line. Front end data entry requires more effort now" (Stedman 1998b).

Data Input as Ease of Use

Despite ERP system requirements that lead to changes in the input data and supporting processes, "ease of use" is a factor that is almost always a part of comparative ERP evaluations. Not surprisingly, some ERP systems reportedly are easier to use than others. System developers are trying to make their ERP systems easier to use, and they have many incentives to do so.

Some ERP Systems Are Easier to Use Than Others

Some systems may be easier to use than others, but ease of use is not the only variable that drives system choice. For example, SunAmerica ultimately chose SAP even though, according to the vice-president of cash services at SunAmerica's annuity unit, "PeopleSoft is very appealing when you look at the screens, while [R/3] looked rigid the way it's delivered out of the box" (Stedman 1998a).

However, others have found that the PeopleSoft system is not easy to use. As noted by Stedman (1998b, p. 24), Algoma Steel has started to use People-Soft's human resources package, but workers still go to the mainframe to get information. One reason is that "[w]orkers have to contend with a dozen PeopleSoft screens, compared with just two or three on the old system."

Similarly, some users argue that particular ERP systems are not intuitively easy to use. For example, the CIO of A-dec (a dental equipment maker) found that, after A-dec installed BAAN's ERP system (Stedman 1998b, p. 24),

- calls to the company help desk increased by 64% (the CIO noted, "that tells you right there that it's not an intuitive application"); and
- the system did not reduce inventory by the amounts that were shipped. Although warehouse employees had entered the necessary inventory data, they had to go to another screen to confirm the transaction before the system actually reduced inventory. Unfortunately, the system did not prompt them to confirm.

Developers Are Trying to Make Systems Easier to Use

Ease of use is important enough that, as reported by Busse (1998), SAP is making efforts to facilitate access to a broader set of users. As noted by Hasso Plattner, CEO and a co-founder of SAP, "I want people to be able to use a portion of [R/3] with zero training." As a result, SAP has begun what it refers to as "EnjoySAP" in an effort to make R/3 easier to use and to make R/3 more adaptable to a wider range of work environments. These efforts likely will employ a broad range of technologies, including touch screens, to facilitate data input.

However, Stedman (1998a) quotes Plattner as follows: "the feedback is that not as many people as we expected are using SAP." This is a critical issue to ERP vendors for a number of reasons. First, if the system is not easy enough to use then it might not be sold to as many firms as would otherwise be the case. Second, even if ease of use does not affect the adoption of the ERP system by particular firms, revenues can still be influenced because they are frequently determined by the number of "seats" sold (i.e., number of system users). As a result, some firms have provided more casual users with access to financial and other information over an intranet, leaving only their expert users with direct access to the system (Bashein, Markus, and Finley 1997).

It is important to note that, to some extent, "ease of use" is in the eye of the beholder. When a system looks like something we have seen before, it may be easier for us to use. For example, Brownlee (1996) noted that one user remarked: "I didn't know how the old system worked .... In my mind, that's probably an advantage."

Summary

Reengineering that derives from ERP systems ultimately results in changes to how many times data is gathered, how data is gathered, where data is gathered, and when data is gathered. Although each change can result in benefits, changes also can generate user resistance. Ultimately, the benefits of the changes need to be weighed against the costs of user resistance, and additional efforts may be needed to mitigate such resistance.

Legacy systems may actually provide more efficient gathering of particular sets of data. For example, what can be gathered in a single screen in a legacy system may take from three to six screens in an ERP system. This provided the basis of a model of implementation difficulty, comparing legacy and ERP systems (a) with respect to the number of screens needed and (b) in light of the number of input transactions. As both factors increase, the difficulty of implementation increases.

Although all ERP vendors recognize the importance of data input capabilities, some systems will probably prove easier to use than others. In any case, ERP developers have initiated efforts (e.g. EnjoySAP) aimed at making their systems easier to use.

References

Bashein, B., Markus, L., and Finley, J. (1997). *Safety Nets: Secrets of Effective Information Technology Controls.* Morristown, NJ: Financial Executives Research Foundation.

Brownlee, L. (1996). "Overhaul." *Wall Street Journal,* November 18, pp. R12, R17.

Busse, T. (1998). "SAP to Broaden Access to R/3." *Computerworld,* September 15.

Hammer, M. (1990). "Reengineering Work: Don't Automate, Obliterate." *Harvard Business Review,* July/August, pp. 104–12.

Koch, C. (1999). "The Most Important Team in History." *CIO Magazine,* October 15.

Stedman, C. (1998a). "R/3 Complexity Stymies Users." *Computerworld,* September 21.

Stedman, C. (1998b). "ERP User Interfaces Drive Workers Nuts." *Computerworld,* November 2, pp. 1, 24.

Whang, S., Gilland, W., and Lee, H. (1995). "Information Flows in Manufacturing under SAP R/3." OIT no. 13, Graduate School of Business, Stanford University, Stanford, CA.

5

ERP Output Capabilities

Access to ERP information was at first limited to the reports that could be generated from the system or through database queries of the underlying database model. Recently, however, organizations have begun to employ intranets and data warehouses to further distribute ERP-generated information. In addition, ERP vendors have begun to develop "portals" that are designed to provide users specific access to a range of information. This has led to improvements in ease of use and hence to new opportunities in electronic commerce for ERP vendors.

ERP Reporting and Query Capabilities

With ERP systems, information is made available to particular users in the form of specific reports. In addition, there are other approaches to generating data from the system, including database queries. Recently, ERP reporting capabilities have begun to evolve as ERP vendors have tried to increase the accessibility and ease of use of the ERP software.

ERP Reports

Enterprise resource planning systems can generate a wide range of standard reports that are designed to meet standard decision-making concerns. The reports that are available depend on the particular module of interest. For example, financial modules produce classic financial reports, including income statements and balance sheets.

However, ERP reports do not always meet the users' needs, as will be seen in a number of examples in this chapter. Consequently, firms may need to find alternative approaches to generating those reports. Note that ERP system reporting capabilities are not easily employed by all users, and there is often a "seat" cost, since ERP system prices are typically based on the number of system users. As a result, some firms generate reports and then put them into other environments (e.g. intranets) so that ERP expertise is not always required to make use of the system's reporting capabilities.

Database Queries

Enterprise resource planning typically sits on top of a relational database. As a result, using either database or ERP query capabilities, reports can be generated and made available based on the underlying information. Queries are particularly helpful in those settings where the reports generated from the ERP do not meet user needs.

Database queries can be done at either of two levels. First, within ERP software there typically is query capability. Second, within the database (e.g. Oracle) encompassed by the ERP software there is also a query capability. Organizations usually find that their expertise lies at one of these levels, and they exploit those opportunities.

Enterprise resource planning systems have been designed to process transactions. Thus, it is probably not surprising that, within some ERP systems, queries are treated as transactions. However, the more queries that are run, the higher the danger that the system will become overloaded; too many queries can lengthen response times and ultimately make the users unhappy. For example, as noted by Koch (1999),

> reports are taking a heavy toll on ERP systems everywhere because for the first time hundreds, even thousands of employees are going to ERP's single, integrated database and pulling out huge piles of data .... This is the number one technology fire that ERP project teams have to put out after the new system goes in.

Thus, a critical part of ERP capacity planning is establishing how output is going to be formulated and implemented. If it is through queries, then additional hardware and network capacity will be necessary.

What If Reporting Capabilities Are Not Sufficient?

If ERP reporting and query capabilities do not meet a firm's needs, then there are at least three alternatives. First, the firm can have some customized reports generated. Second, the firm can put the report information on an intranet. Third, a data warehouse could be developed for extended and ad hoc queries.

Customized Reports

Customized reports offer the advantage of seemingly being able to get specific decision makers exactly what they want in a report that is produced periodically.

However, there are a number of disadvantages to be offset against that apparent advantage. First, decision makers' needs change over time in response to changes in their environment and in their understanding of the problems they are solving. Second, specific decision makers in organizational positions change: users are promoted, hired, and fired. Given that a report is specifically designed for a particular user, if the user changes then the report may also need to change. Third, specifically designed reports may reach for information that is outside the scope of the decision maker or may not include all of the relevant information. Fourth, customizing the ERP can be costly and can restrict the ability to upgrade the system. In sum, building specific, customized reports into the system can be a costly and very short-run approach.

Intranets

Because ERP systems are not, in general, easy to use, some firms have pursued policies of placing reports on the World Wide Web or an intranet. In particular, different reporting environments can be generated for different levels of users. For example, expert users could be given direct SAP access while occasional and casual users are given access to the information they need, in an easy-to-use format, on the corporate intranet.

Increasingly, intranet and ERP integration is becoming a vehicle for ERP report information. For example, Transamerica in Los Angeles uses an intranet to make report information available from its PeopleSoft system. Reports, developed to meet both general and specific decision-making interests, are updated every 24 hours.

Moreover, the well-known Lotus Notes can also be used to access ERP data in order to generate reports that are easier to read and use. For example, Lotus Notes allows access to SAP and PeopleSoft data (Doan 1996). Such access allows organizations to make ERP data available in a system that users are familiar with and that is easy to access and use.

Reports published on an intranet are typically standard reports that are "spun off" for accessibility. As a result, the information itself is generally about the same as that available from the standard ERP reports. What changes is the availability: now the information is generally available to a broader range of users.

Data Warehouses

In some instances, ERP systems are interfaced with data warehouses. Data warehouses can provide at least two capabilities. First, they allow the user the ability to access decision-making information in an environment designed to facilitate the generation of reports. Second, the data warehouse can function as a sort of clearing house for upstream or downstream applications to both internal and external users. For example (Cotteleer, Austin, and Nolan 1998, p. 10), Cisco used a data warehouse in their Oracle Applications implementation:

> Because the downstream impacts of the [Oracle Applications] project were so much greater than expected, the team decided to tackle some larger technical issues. Whereas before systems had tended to communicate directly with one another (i.e., "point to point"), a new approach would now be employed in which all data communication would take place via a "data warehouse."

Data warehouses may be located in a single physical location or they may be "virtual" data warehouses, with parts of the database stored in multiple locations. Appendix 5-1 describes how Quantum's ERP is interfaced with a virtual data warehouse. As described next, Burton Snowboards employed a physically distinct data warehouse.

Data Warehouse at Burton Snowboards (Kempster 1998)

Burton uses SAP's R/3 as a transaction processing system to capture data from its three offices in Austria, Japan, and Vermont. Burton found that generating nonstandard reports from SAP was too difficult. The small information systems group could not accommodate the requests because of SAP's complexity. In addition, Burton found that even some routine tasks were difficult to accomplish using SAP's reporting capabilities. For example, Burton's MIS director noted that

> We also needed something that would allow an inventory manager or product manager the ability to track inventory without running a reporting request through the IS department .... A wide range of people need this information, like our sales, inventory and operations teams and our IS group.

In order to accommodate both routine and nonroutine inquiries, Burton implemented a data warehouse. Burton "migrates" information daily from the SAP system to its data warehouse and then uses a management reporting tool (Vista/EA) to view the data. Roughly 70% of Burton's middle managers, and almost all of their executives, use output from the system.

Knowledge Discovery

A data warehouse can be used to provide more than just reports. Given all the data in a single location, firms have the ability to generate or discover knowledge in the data. For example, relationships between different variables can be investigated using statistical and artificial intelligence approaches. Such knowledge allows value creation from transaction processing activities. Previously, costs of transaction processing systems were viewed as purely overhead.

Table 5.1. *ERP Vendor Portals*

Vendor	Portal
J.D. Edwards	MyActivEra
Lawson	Insight II Seaport
PeopleSoft	MyWorld
SAP	MySAP

Advantages of Data Warehouses

Increasingly, ERP vendors are integrating data warehouse capabilities into their systems. Because of their focus on databases, Oracle was perhaps the first major vendor to integrate data warehouses into their ERP product. However, SAP was not far behind with its own data warehouse product.

Data warehouses offer the availability of information that is optimized for retrieval. Thus, as decision-making needs change, the data warehouse can be used to provide new reports in a real-time manner. As seen in the Cisco example, data warehouses can also be used as a central interface point. The costs are basically those of generating and maintaining the data warehouse and the cost of writing the data warehouse queries.

ERP Portals

As firms have increasingly turned to intranets and data warehouses, ERP vendors have begun to focus on providing portals that facilitate development of custom desktop access for ERP users. Some of the ERP vendor portals are summarized in Table 5.1.

Among ERP vendors, the notion of a "portal" has not completely stabilized. As noted by an analyst at the META Group: "These vendors are throwing things up against the wall – a dozen different things – and seeing what sticks" (Holt and Lamonica 1999). In general, an ERP portal is a website (on an intranet or the Internet) that provides access to news and information directly or indirectly related to the ERP system. Enterprise resource planning portals provide the opportunity to sell related products in an electronic commerce environment. The portals of some ERP vendors (e.g., SAP's MySAP.com) allow other vendors with SAP-related products to sell their goods or services.

Two versions are used: desktop-based and web server–based. Lawson recently made available a sample custom desktop design (see Figure 5.1) aimed at a specific worker or job function. As part of the design, the user would have access to different ERP submodules of concern (e.g., requisitions, purchase orders), a help-desk function, and different reports. Electronic commerce is

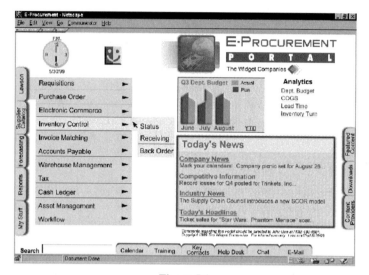

Figure 5.1

embedded on the same page through access to supplier catalogs, Lawson's pages, and content providers. In addition, the page provides a place to centralize issues of personal concern, such as calendar, key contacts, e-mail, and so forth. Finally, the desktop has an area for news that can be used by the company to direct appropriate information to employees.

MySAP is a web server–based portal (see Figures 5.2 and 5.3) that is available for personalization. Users personalize their portal access to *topics* (business directory, business processes, industry trends, etc.) and *channels* (e.g., industry focuses such as high tech).

To the extent that intranets and data warehouses signaled a loss of control of some ERP information, the use of portal-based custom desktops and personalized web server pages begins to bring back some control to ERP vendors. Whereas shifting information to a general corporate intranet pushed the information out of the ERP system, portals can now centralize information within the context of vendor products yet still allow more general access.

Summary

Generating reports from an ERP system can be accomplished using a number of different approaches, including database queries and ERP reporting capabilities. However, some firms have made information available to more casual users via their corporate intranets. In addition, data warehouses provide an important vehicle for generating reports or capturing views. One additional advantage of

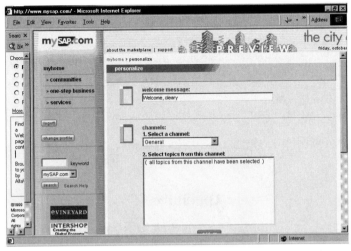

Figure 5.2

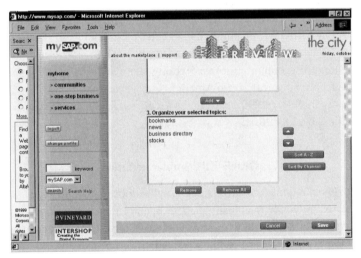

Figure 5.3

the data warehouse approach is that knowledge discovery approaches can be used to tease knowledge out of data generated as part of the ERP systems.

Recently, the focus of output presentation in ERP systems has shifted away from a report focus to a custom-designed desktop that employs portal information. In addition to report access, these role-specific designs allow access to the appropriate portions of the ERP system and its reports, news items, and personal items such as calendars and e-mail.

References

Cotteleer, M., Austin, R., and Nolan, R. (1998). "Cisco Systems, Inc.: Implementing ERP." Report no. 9-699-022, Harvard Business School, Cambridge, MA.

Doan, A. (1996). "Domino Gets Back-End Links." *InfoWorld,* December 2.

Holt, S., and Lamonica, M. (1999). "ERP Vendors Join the Ranks of Web Portal Sites." *InfoWorld,* May 31.

Kempster, L. (1998). "ERP, Warehouse Used in Concert." *Computerworld,* September 21.

Koch, C. (1999). "The Most Important Team in History." *CIO Magazine,* October 15.

Appendix 5-1

Quantum's Virtual Data Warehouse

Quantum

Quantum Corporation was founded in 1980. A diversified mass storage company, Quantum manufactures hard disk drives for personal computers and all classes of tape drives. Quantum is a market leader in both fixed and removable storage. Quantum sells storage products to original equipment manufacturers (OEMs) and distribution customers worldwide. Quantum has nine business operations located across North America, Europe, and the Pacific Rim. In the two-year span 1996 to 1998, sales for the fiscal year ending in March grew from $4.4 billion to $5.8 billion.

Oracle ERP Implementation

In 1996, Quantum implemented Oracle Applications, in both financial and production, on an Oracle relational database management system. The resulting system had over 40,000 underlying tables. Unfortunately, this huge number of tables made it difficult to use the information for decision-making purposes. The back-end transaction processing environment was too complex for many users, and the system seemed destined to be a repository of largely unused

---

This appendix is based on information from the following sources: "Quantum Corporation Creates a 'Virtual Data Warehouse' to Win 1998 Best Practices Award from the Data Warehousing Institute," ⟨www.quantum.com/corporate/pr/dataware.html⟩; "Category: Architecture Implementation – Meta Data Management, Winner: Quantum Corporation," ⟨www.dw-institute.com⟩; "Brio Technology and Integral Results Work Together to Ensure Quantum's Success," ⟨www.brio.com/customers/ss.quan.html⟩; and "Oracle Enhances Discoverer 3.0 Analysis of Oracle Applications Data," ⟨www.noetix.com/o/newsroom/pr_disc2k.htm⟩.

transaction information. As a result, Quantum saw that they must take steps to make the information more easily accessible.

Virtual Data Warehouse

In order to make the transaction information available to managers and other decision makers, Quantum decided on a data warehouse. About four thousand simplified and business-oriented database views were created to simplify access to data. Kevin Conway, information services (IS) program manager, observed:

> Our goal in Quantum's IS department is to empower our end users to access corporate data to run the business.

> Our challenge was to make all transaction and operations data from our Oracle enterprise resource planning system available to end users as working business information – and to do this globally with a number of different business units.

Quantum's virtual data warehouse provides a "next day" look at all of Quantum's transactions:

> Quantum's information services department created what has been termed a "virtual data warehouse" – a unique strategy combining several different technologies allowing Quantum's nine global business operations to meet crucial reporting requirements by providing on-demand available reporting for all of the company's transactions. This multi-server strategy allows the preservation of continuous robust on-line transaction processing while also allowing rapid business reporting: neither business function is hindered by the other and work can be completed in much less time.

Data Warehouse Use

Currently, the data warehouse averages 250 users (during any shift) 23 hours per day, six days a week. Users come from business units worldwide. The system was designed to meet the business intelligence needs of Quantum's functional users, including transportation, finance, sales and operations, and operational and tactical managers.

Diana Chan, a manager of strategic process management, observed that

> [t]his system allows us real-time access to data in consolidated reports, which streamlines data processes. This real-time access allows us to make smarter business decisions. We can now track where products are in transport, if suppliers are providing deliveries on time, and vendor price differences that we can pass on to our buyers and managers immediately. All of these help us reduce costs and serve our customers and partners more effectively.

The data warehouse has allowed integration and "fine tuning" of some reports in order to meet decision-maker needs in a more rapid manner. As one example, what is now provided in a single report formerly required fifteen different reports and sixteen hours to construct.

Implementation Firms

Development of the data warehouse required integrating the work of a team of different firms in addition to Oracle, including Hewlett-Packard, Brio Technology, Noetix Corporation, and Integral Results.

- The data warehouse was implemented in a client server environment using Hewlett-Packard HP 9000 Enterprise Servers.
- Noetix Corporation developed Noetix Views, which provides a comprehensive set of business views designed for Oracle Applications modules, including general ledger, accounts payable, accounts receivable, inventory, purchase order, and order entry.
- Brio Technology provided on-line analytic processing (OLAP) and reporting capabilities to users in client server and World Wide Web environments. Quantum uses Brio Enterprise in conjunction with Noetix Views to access the data from the Oracle Applications. Also, Oracle's data warehouse team uses BrioQuery to generate new queries.
- Integral Results is a consulting firm that specializes in data warehouses and decision support systems. Integral Results taught users how to query and view data with the BrioQuery tool. Integral Results has partnered with Brio Technology since 1993 on a range of projects. Brock Alston (western regional manager for Brio Technology) recommended Integral Results to help Quantum customize their training for Brio's products. As noted by Kevin Conway, "to be successful, we had to have our customers and our drive discs model numbers within the training examples."

Quantum's Role in the Implementation

Quantum used roughly a 50/50 mix of information services and business user resources to implement the system.

> What made the implementation work was our global teamwork .... Business systems analysts at each site led the way by understanding user operational information requirements, building queries, and teaching classes. Regional information systems programmers and analysts provided technical assistance as the need arose. This teamwork allowed global sites to learn from each other and

made the implementation proceed more quickly and with a higher degree of user acceptance.

Data Warehouse Staffing

The data warehouse requires a team to keep up with current users' needs, system changes, and training requirements. In addition to the program manager, the data warehouse is staffed by two consultants (senior programmer analysts), two information service business analysts, one technical trainer, and one desktop/customer support installer. As noted by Loren Gruner, "Kevin Conway ... has built a whole team that supports Quantum's global infrastructure."

Emerging Issues

As part of the implementation of the data warehouse, Quantum faced a number of challenges: world-wide training; developing expertise in data access; using multiple hardware, software, and consulting firms; and maintaining project momentum.

Worldwide Training. Because the data warehouse was being installed on a worldwide basis, training would likewise have to be supported on a worldwide basis. As a result, there would be problems with language and other issues.

Developing Expertise in Data Access. With the development of the data warehouse there was now a need for Quantum to develop an expertise in data access and reporting in order to meet ad hoc and on-going reporting requirements. This required staffing, training, and tools.

Multiple Hardware, Software, and Consulting Firms. Integrating the efforts of multiple firms is never easy. However, Loren Gruner (president of Integral Results) believes that the project teams involved each had the same overall mission:

> I think that both companies [Brio and Integral Results] share a philosphy of customer service. Customer service means understanding what it takes to get a good product – whether you are shipping product or implementing a product at a customer's site.

Questions

1. Why did Quantum use a data warehouse when it already had an enterprise resource planning system?
2. Why did Quantum use a 50/50 mix of information services and business users?

3. What is a virtual data warehouse? How do you think it might differ from a regular data warehouse?
4. What are some of the costs of developing and running a data warehouse? How effective is the cost–benefit trade-off?
5. What kind of partnerships do we see in the implementation of data warehouses?

6

Technology Enabled versus
Clean Slate Reengineering

The primary tools for capturing and implementing best practices are enterprise resource planning (ERP) systems. As noted by Hammer (1997), "SAP implementation equals forced reengineering." A vice-president of Cap Gemini remarked that, regarding SAP implementations, "it's rare when you don't have to do some kind of reengineering" (Gendron 1996). Gendron goes on to call ERP "the electronic embodiment of reengineering." Because reengineering is so tightly tied to ERP implementation, firms must address the issue of using clean slate versus technology enabled reengineering.

The purpose of this chapter is to investigate the relationship between these two types of reengineering. In particular, this chapter elicits some of the advantages and disadvantages of each approach, focusing on why some firms have chosen one or the other approach as part of their reengineering efforts. In addition, this chapter investigates some of the firm characteristics that are likely to lead to using a particular reengineering approach.

Reengineering Tools and Technologies

Hammer's (1990) ideas for reengineering were initiated before there were tools to facilitate reengineering. Many early reengineering efforts were designed to "obliterate" existing processes, but there were few tools to guide firms as to how the processes should be reengineered. As a result, one of the primary reasons that many reengineering efforts failed was the lack of a sufficient tool set to facilitate reengineering. An early analysis (CSC Index 1994) of tools that were used in reengineering focused on process value analysis, benchmarking, competitive analysis, and activity-based costing. Table 6.1 summarizes CSC Index's (1994) study, which was designed to understand what "technologies" were used to help in the process of choosing which process or artifacts needed to be reengineered.

Table 6.1. *Tools Used to Choose*
What to Reengineer

Tool	USA	Europe
None	41%	36%
Process value analysis	36	27
Benchmarking	34	36
Competitive analysis	25	28
Activity-based cost analysis	20	17
Other	16	17

Source: CSC Index (1994).

Because this survey preceded the broad-based introduction of ERP systems, ERP was not one of the tool options. Perhaps the closest is "benchmarking," where best practices of other firms could be captured. In addition, since the time of this original survey there have been other tools developed to facilitate reengineering, some of which have been embedded within the context of ERP systems. Currently, however, ERP is the dominant facilitating technology for reengineering. But with any new tools come the same concerns of technology enabled versus clean slate reengineering.

What Are Technology Enabled and Clean Slate Reengineering?

Reengineering is actually implemented along a spectrum of approaches that range from a technology enabled approach to a clean slate approach.

Technology Enabled (Constrained) Reengineering

In technology enabled reengineering, a particular technology (or portfolio of technologies) is chosen to perform the reengineering before any systems, artifacts, or processes are reengineered. As a result, reengineering choices are a direct function of the particular technology chosen to perform the reengineering. For example, if the technology is an ERP system, first an ERP system is chosen; second, the enterprise MAPs are chosen from those that the ERP software can accommodate. Hence, the ERP software drives the reengineering. The ERP software can accommodate all the necessary basic documents (virtual or paper) in the chosen design, since the design is derived from the software. As a result, the reengineering is said to be "technology enabled." Since the implementation is not altering the software, the time and cost of implementation are likely to be lower with technology enabled implementation.

Technology enabled reengineering has been referred to by other names. Gemini Consulting (1996, p. 7) labels the simultaneous reengineering and implementation of an ERP as "concurrent transformation." In addition, in the same sense that the technology can be said to enable reengineering, the technology also could be said to *constrain* the reengineering to the set of options that can be accommodated by the ERP system. As a result, we will also refer to this as "constrained" reengineering.

Clean Slate Reengineering

In clean slate reengineering, the system design starts with a "clean slate." Processes are reengineered to meet enterprise needs and requirements. The clean slate approach might also be referred to as the approach of "starting from scratch." Using this approach, theoretically there are no predefined limiting artifacts or processes. Ideally, developers can generate an optimal system design for the particular organization. In the case of ERP systems, this approach implies a sequential ordering of reengineering first and then choosing the ERP software. The software is thus "fit" to the reengineered organization – that is, the software is fit to meet the needs of the clean slate design. Since the software is customized to meet the needs of the design, implementation costs of clean slate generally exceed those of technology enabled reengineering. However, with a clean slate approach the new system should better meet the organization's needs.

Somewhere in Between

In most reengineering projects, the choice of clean slate or technology enabled is not a discrete one; instead, there is a continuum between the two. Typically, reengineering occurs somewhere on that continuum – either more toward clean slate or more toward technology enabled, employing greater or lesser portions of each. These issues are important, and we return to them in Chapter 9, where we focus on which should be changed: the software or the organization.

Advantages of Technology Enabled Reengineering

There are a number of advantages of technology enabled reengineering, each of which derives from exploitation of the capabilities of the technologies used to facilitate reengineering. The case of Geneva Steel is useful for illustrating some of the advantages of using ERP as a tool for reengineering.

ERP Provides an Ideal Goal for Reengineering

In the Geneva Steel case, the firm adopted and used SAP as a tool for changing information flows, restructuring the firm. The ERP system was seen as a technology that would facilitate change in the organization by integrating databases, improving financial reporting, reducing the number of people, and so forth.

Tools Help Structure Complex Reengineering Efforts

Particularly in the case of companies with complex processes, tools that facilitate reengineering (e.g., ABC or ERP) can provide a map or a guide to process and artifact development. For example, ERP systems provide a guide as to what activities are involved in what processes, how artifacts need to be structured, and so on. Enterprise resource planning systems provide users with both a starting point and a completion point. As a result, tools like ERP systems provide a structure that may be critical to the success of a particular reengineering effort.

Tools Help Rationalize and Explain Reengineering Efforts

When used as reengineering tools, ERP systems also provide a "reason" for reengineering: "We have to reengineer in order to implement SAP." Geneva Steel made it clear that it would be adopting SAP and that this would change the organization: "people have now finally figured out that whether you didn't like or whether they liked it, it [was] going to happen." The SAP implementation was seen as providing the new organizational map.

Tools Help Build Better Solutions Than Would Otherwise Be Built

Enterprise resource planning systems can guide the reengineering effort to help produce better solutions than might otherwise be chosen. This is most likely to be the case if experience is required to build better solutions and the development team has limited experience. In effect, the best practices embedded in ERP systems represent implementable knowledge about processes that could mitigate problems of inexperience. In the case of Geneva Steel, SAP had been used by a large number of steel firms. As a result, the best practices of many steel firms had been embedded in the software prior to Geneva's adoption of SAP's ERP software. Accordingly, Geneva Steel was in a position to find better solutions than it might have discovered on its own.

Technology Enabled Reengineering Puts "Bounds" on the Design

Technology enabled reengineering limits the set of design choices. If there are no bounds on the design then there can be a large number of designs, resulting in information overload and difficulty in choosing between particular designs. Without bounds, processes can begin to merge into each other, making process boundaries unclear. For example, process boundaries can provide guidelines for resource use, personnel involvement, and project management.

The Design Chosen Is Feasible for the Software

Perhaps the most important reason for using the technology enabled approach to reengineering is that designs are chosen from a portfolio of feasible designs. The ERP system is designed to accommodate each set of design choices. Further, choosing from a given set of design choices provides process designs that can interact with other process designs chosen for use in the system. Thus, feasibility is critical to particular processes and their interaction with other processes. For instance, Geneva would be making its choice of best practices from the SAP portfolio, which meant that the design chosen would be feasible.

There Is Evidence That the Design Will Work in an Organization

If a best practice process is part of an ERP or other knowledge base of best practices, then there is evidence that the process does in fact work. That a design is included as a best practice typically indicates that some organization, somewhere, has successfully implemented this design. This is one of the primary reasons that Nestlé chose to implement a portfolio of best practice processes and artifacts that was generated from a consultant's database of best practices and the ERP company's best practices. Similarly, since SAP had been used in a number of steel companies, there was substantial evidence that SAP would work in the case of Geneva Steel.

Designs Are Likely to Be Cost-Effective

Enterprise resource planning software developers have become sensitive to the notion that, when their software is used, it is desirable to have the implementation cost be reasonable. For example, SAP's recent development of "ASAP" has focused on getting SAP implementations up and working in a cost-effective manner. In general, system costs related to implementing existing designs (e.g., best practices in an ERP system) are likely to be reasonable.

Designs Are Likely to Be Implemented in a Timely Manner

Developers have also become aware of the importance of being able to implement ERP software in a timely manner. As just noted, SAP has developed ASAP, a design approach aimed at keeping implementations to a reasonable length of time (e.g., nine months). As a result, most design choices provided by the ERP software can be implemented in a timely manner.

Avoid Multiple Layers and Waves of Consultants

A discussion with a Belgian company found that they used SAP enabled reengineering because they were not pleased with their previous experiences of consultants generating a number of suggestions and accompanying paperwork. These suggestions had not been implemented for three primary reasons. First, although the ideas were written down in report form, they could not be easily generated into a form that could be actually used in a system (ERP-based best practices can be directly implemented). Second, the feasibility of suggestions was difficult to evaluate; and finally, generation of the ideas was expensive.

Software Is Available

Reengineering that is ERP technology enabled is guaranteed to have a software solution available. This is not the case for clean slate reengineering, where the designs can be developed but there may be no software available that meets the requirements of the design.

Advantages of Clean Slate Reengineering

Clean slate reengineering has a number of advantages over technology enabled (e.g. ERP-based) reengineering that result from its lack of dependence on any particular technology. The case of Geneva Steel's Delta Project (see Appendix 3-1) provides an example of clean slate reengineering.

Not Constrained by the Limitations of Any Particular Tools

Clean slate reengineering is not constrained by the limitations or biases of any particular tools. For example, Geneva Steel's Delta Project did not have any initial biases as part of its implementation, since there were no tools to facilitate reengineering. Clean slate reengineering can thus encourage use of a portfolio of tools to facilitate implementation. Since any single reengineering tool has limitations and biases, a portfolio of tools could have substantial advantages over a single tool.

Not Constrained by Knowledge about Artifacts and Processes

Whereas ERP systems can employ a range of best practices, they are also limited by their database of best practices. Firms that have unique value-creating processes often work to make sure that they are *not* included in the portfolio of processes that go into consultant databases or ERP systems. Discussions with Price Waterhouse led to the finding that a number of their clients requested that information about their best practices not be included in the best practices database. Some clients have even asked Price Waterhouse to exclude publicly available knowledge from those databases. As a result, whenever an organization uses a best practices database, it may not include all the available best practices.

Future Versions Are Not Necessarily Limited by Changes in a Particular Technology

Inevitably, processes evolve over time. That evolution is based on a number of factors, including the starting point. In general, the better that starting point, the less that the process will need to change over time. If a design is technology enabled and the technology changes, this can put the technology enabled reengineering firm at a disadvantage. Users of ERP software effectively "buy in" to future versions of the software if they want their system to evolve. Otherwise, there is a large set-up cost with movement to an alternative system.

Developing a Design to Which Others Do Not Have Ready Access

Clean slate approaches allow firms to think "out of the box." In so doing, firms can generate new approaches. This is particularly important in settings where reengineering can provide a competitive advantage (rather than facilitate catching up). For those firms where technology is used as a competitive edge, clean slate reengineering can provide an additional advantage. With clean slate reengineering, only the developing firm has knowledge of the design. If existing best practices are used instead, firms are often just playing "catch-up."

Reengineering Is Treated Separately from Technology Implementation

Oftentimes ERP implementations have been criticized as being too expensive and taking too much time. In some situations these drawbacks arise not from the ERP software implementation per se but rather because the ERP included both software implementation and reengineering. With clean slate reengineering, there is no confusion over which is the reengineering and which is the technology implementation.

Gemini Consulting's (1996) survey on business process reengineering (BPR) and ERP found that only 16% of firms surveyed indicated that they had planned to pursue BPR before ERP implementation. After implementation, however,

35% said that BRP before ERP implementation was the strategy to pursue. As a result, there is some evidence that firms need to pursue reengineering first and then ERP implementation.

Clean Slate May Be the Only Way for Processes Embedded
in New Technology

In some situations, processes must be embedded in the context of new technology – for example, the World Wide Web or barcode scanning – when they are first available. There may be little or no understanding of what is a best practice when using such technologies. Since there is limited understanding of how the supporting processes are expected to work in the context of new technologies, clean slate reengineering may be the only feasible approach in such cases.

Disadvantages of Technology Enabled Reengineering

The disadvantages of technology enabled reengineering are largely the reverse of the advantages of the clean slate approach. Since the latter have already been discussed in some detail, the former are simply bulleted here.

- Reengineering is limited by the particular tool used for the implementation.
- Reengineering is limited by knowledge about artifacts and processes that are embedded in the tools.
- Evolution of the system may be limited by the technology.
- The design chosen is one that other firms have access to.
- There may be some confusion as to whether it is a technology implementation or a reengineering activity.
- There may be no best practices available for some settings, limiting its use.

Disadvantages of Clean Slate Reengineering

Disadvantages of the clean slate approach are essentially the reverse of the advantages of the technology enabled approach. Since these have already been detailed, a bulleted list will suffice here.

- There may be no structure to help with the design.
- There is no rationale for reengineering.
- The designs may be suboptimal.

- There are no bounds on the initial design.
- A chosen design may not be feasible.
- A design might not work with the chosen ERP software.
- It may be costly and time-consuming to develop and implement the chosen design.
- There can be many waves of consultants and layers of reports summarizing their work.
- There may not be software available.

Who Should (or Does) Use the Clean Slate Approach?

Which firms should use a clean slate approach and which should use a technology enabled approach? The answer depends on a number of factors, including firm size, whether or not the firm has a lot of money to spend on development of new processes, whether or not the firm has enough time to generate processes, the extent to which a firm depends on the technology to generate their competitive advantage, and how unique are the processes used at the firm.

Large Firms

There is a size bias in using the clean slate approach, with larger firms having a number of reasons for pursuing this strategy. Large firms:

- have the resources to do clean slate reengineering;
- often are industry leaders and, as a result, are likely to have more time;
- are more likely to use processes as a basis of strategic advantage; and
- are more likely to need a unique solution.

Firms with "Deep Pockets"

Clean slate reengineering requires substantial resources, because the reengineering effort is not exploiting the capabilities or information that could be provided by the ERP best practices. In addition, in order to implement clean slate reengineering, the firm must have resources available to try to generate new processes and artifacts. Innovation also requires involvement by particularly high-quality employees who might otherwise be pursuing other corporate activities.

Procter & Gamble illustrated the importance of deep pockets with their work on reengineering supply chain management processes (see McKenney and Clark 1995). Procter & Gamble used a number of different designs in a number of different client companies in order to develop their supply chain management system, which they eventually gave to IBM. Starting with electronic data interchange, Procter & Gamble evolved toward a system where they monitored customer inventories and placed orders for customers using direct connection to customer databases. The evolution required substantial time and investment, which would not have been possible without substantial resources.

Firms with Time

The amount of time that a firm has available to implement an ERP can depend on a number of factors, including the firm's motives for adopting the ERP. In the case of Quantum, an ERP model was chosen to be implemented in order to facilitate competition on the customer service front. Customers could not wait for Quantum to respond to order requests, and Quantum was therefore losing business. Computer disk drives had become a commodity, so Quantum was interested in pursuing customer service as its competitive advantage. It needed a way to integrate its factories and suppliers around the world; hence, they had limited time to spend on clean slate reengineering. As a result, they employed ERP technology enabled reengineering.

Firms for Which Processes Are Used to Create Strategic Advantage

Firms can generate competitive advantage either by generating better processes or by implementing existing processes better than other firms. If firms use existing processes, whether or not they are best practices, then ultimately they are choosing to compete based on implementation of that portfolio of processes.

  The more unique a firm in terms of its industry, its processes, its customers, or any of a number of other factors, the more likely it is they will *not* be interested in using the processes or artifacts of other companies that would be available through technology enabled reengineering. Instead, in order to accommodate their unique processes, these firms may find it necessary to use clean slate reengineering.

Firms That Seek a Unique Solution

Technology enabled solutions can be copied by others, and ERP best practices implemented at one firm can be implemented at others. In contrast, clean slate reengineering solutions are not usually disseminated as rapidly. If an organization needs a unique solution, then clean slate reengineering is more likely to provide one.

Who Should (or Does) Use the Technology Enabled Approach?

The factors just discussed can also be used to determine which firms should use the technology enabled approach. For example, smaller firms typically

Table 6.2. *Business Process Reengineering
and SAP R/3*

Panel A: Original implementation strategy	
BPR and SAP R/3 implementation simultaneously	48%
BPR before SAP R/3	16%
BPR after SAP R/3	3%
BPR before and after SAP R/3	1%
No BPR needed	33%
Panel B: After implementation, which would be selected?	
BPR and SAP R/3 implementation simultaneously	51%
BPR before SAP R/3	35%
BPR after SAP R/3	3%
BPR before and after SAP R/3	1%
No BPR needed	10%

Source: Gemini Consulting (1996).

have small budgets and standard processes that limit their ability and need to do much clean slate reengineering. Accugraph (see Koch 1996) implemented SAP's prestructured "Special Delivery" package; they focused on the financial package, consultants, and no reengineering outside the scope of the prespecified design.

Which Approach Is Used the Most?

Gemini Consulting (1996) gathered empirical information on simultaneous versus separate BPR (business process reengineering) and SAP implementation. They asked firms for (a) the actual strategy used and (b) the preferred strategy, were they to do it over again. Although the data concerns SAP specifically, it is likely that the results can be generalized to other ERP systems. The results of that survey are summarized in Table 6.2.

Gemini Consulting found that, prior to or after implementation, only 4% would plan or do BPR, after or before *and* after, SAP implementation. Instead, 96% would either do no BPR, do BPR before SAP, or do BPR simultaneously with the SAP implementation. Gemini Consulting (1996) found that only 16% had planned reengineering before the SAP implementation; afterwards, 35% said that they would pursue that strategy. In both panels A and B, technology enabled ERP implementation is the dominant strategy. The largest increase from panel A to panel B occurred with the clean slate approach. The largest decrease (33% to 10%) occurred with firms thinking that no BPR was needed.

Accordingly, it seems that firms should consider reengineering when implementing an ERP such as R/3.

Summary

This chapter has investigated clean slate and technology enabled reengineering. Advantages of clean slate reengineering include the following.

- Firms are not constrained by limitations of particular tools.
- Firms are not constrained by a lack of knowledge of best practices.
- Future versions are not limited by the technology.
- Firms can develop unique process designs.
- Reengineering is not confused with technology implementation.
- It may be the only way to build processes for use in new technologies.

On the other hand, technology enabled reengineering can help

- structure the reengineering efforts,
- rationalize and explain reengineering efforts,
- build better solutions,
- bound the design efforts,
- ensure feasible designs,
- ensure that designs work,
- generate more cost-effective design processes,
- ensure timely implementation, and
- firms avoid multiple layers and waves of consultants.

Enterprises should choose the approach that meets their needs and resource constraints in the reengineering process. In general, we can anticipate that firms pursuing a clean slate approach, in addition to being larger, will also have deep pockets, will not be time-constrained, and will tend to have or use unique processes as a basis of strategic advantage. Conversely, technology enabled reengineering is more characteristic of

- firms with limited budgets,
- time-constrained firms, and
- firms that have relatively standard processes.

In any case, firms should recognize that some reengineering will be necessary. In one survey, a majority of firms chose simultaneous implementation of reengineering and ERP – that is, a technology enabled approach to reengineering.

References

CSC Index (1994). *State of Reengineering Report.* Boston: CSC Index.

Gemini Consulting (1996). "Business Leader's Experience with SAP Implementation." Gemini Consulting, Hamburg, Germany.

Gendron, M. (1996). "Learning to Live with the Electronic Embodiment of Reengineering." *Harvard Management Update,* November, pp. 3–4.

Hammer, M. (1990). "Reengineering Work: Don't Automate, Obliterate." *Harvard Business Review,* July/August, pp. 104–12.

Hammer, M. (1997). "Reengineering, SAP and Business Processes." Unpublished presentation given at SAPphire (Orlando, FL), August.

Koch, C. (1996). "Flipping the Switch." *CIO Magazine,* June 15.

McKenney, J., and Clark, T. (1995). "Procter & Gamble: Improving Consumer Value through Process Redesign." Report no. 9-195-126, Harvard Business School, Cambridge, MA.

PART THREE

ERP LIFE CYCLE

7

Deciding to Go ERP

This chapter analyzes some of the different rationales associated with deciding to go with ERP. In particular, this chapter examines four questions.

(1) What are the business case rationales used for determining whether or not to go ERP, and which are most frequently used?
(2) How can business case rationales be used to facilitate design or measure success?
(3) How do firms measure rationales: in monetary or nonmonetary terms?
(4) How does a firm ultimately decide to go ERP?

Business Case Rationales

Business case rationales for ERP systems can be grouped into four different sets of categories: technology, business process, strategic, and competitive. Each of those rationales has different strengths and limitations. This chapter illustrates rationales from each category.

The choice of business case rationale is important for at least three reasons. First, critical analysis is necessary to ensure that the firm makes the correct decision regarding ERP. Ultimately, that analysis can generate a number of different rationales or arguments used to substantiate ERP adaptation. Second, in some cases, the basis of choice can facilitate ERP design, providing design specificity. For example, if the goal is to improve a specific process then the extent of that improvement can be measured. Third, the basis of choice can facilitate evaluation of the success of the implementation, based on whether the business case rationale was successfully met. Different measurement approaches can provide a basis for determining success.

Table 7.1. *Technology Motivations*

Motivation	Number of Firms	Percent of Sample
Systems not Y2K compliant	42	27
Disparate systems	37	24
Poor-quality systems/visibility of information	26	17
Business processes or systems not integrated	19	12
Difficult to integrate acquisitions	12	8
Obsolete systems	11	7
Unable to support growth	8	5

Technology Rationales

Organizations have used a number of technology rationales to justify the choice of an ERP system. A recent survey by Deloitte & Touche asked firms to specify why they decided to go with ERP. The results from that survey are summarized in Table 7.1, where firms could specify more than a single motivation or benefit. In this section we examine in greater specificity a number of the technology rationales, using case studies to illustrate this motivation.

Year 2000 Issue

One of the most frequent reasons for adoption of ERP systems has been the year 2000 issue, also known as "Y2K." Firms were faced with updating legacy programs in order to eliminate potential Y2K difficulties. In many cases this proved to be an almost intractable problem. Since programmers had retired and programs were poorly documented, little knowledge remained regarding many of those systems. Furthermore, there were hundreds of programs that were affected.

Enterprise resource planning appeared to offer a relatively quick and easy solution to many firms. Packages such as SAP were known to be Y2K compliant, so many Y2K problems would be eliminated if particular ERP systems were implemented. In many cases, ERP offered a cost–beneficial solution to the Y2K problem. By going with an ERP system, firms avoided spending millions of dollars without making any improvements over existing solutions.

Case Study. A conversation with Barry Seid, ERP deputy director of Litton Data Systems, yielded the finding that Y2K was the driving force in the sale of the notion of ERP to management. As Seid noted,

We really sold [ERP] on the year 2000 .... If you have systems that are 20, 25, or 30 years old, the Gartner Group ... has indicated that it will cost you anywhere from $1.10 to $1.65 per line of code to change for year 2K. If you have 4,000,000 lines of code you are talking about a lot of money. Additionally, the legacy systems are not there, there is nobody there to maintain them and there is nobody who understands them. So if you had to fix it up for the year 2K it would take you millions and millions of dollars, with a terrific risk. In essence we sold this system on a year 2K basis.

Disparate Systems

Disparate computing environments limit the ability of firms to integrate the different business units that are supported by those various computers. In contrast, ERP systems provide a wide range of software within a single computing environment.

Moreover, many firms made the move to client server computing to break away from the monolithic mainframe and supporting technologies (see e.g. Vaughn 1996). As they made this shift, they also needed new client server software. However, once firms changed to client server computing in the early 1990s, they found that there was little software available for this environment. The limited software that *was* available included ERP software, such as SAP's R/3.

Case Studies. Geneva Steel's information system configuration had evolved over time into a number of heterogeneous systems. As noted by Joseph Cannon (Geneva's CEO),

we have ... a mainframe ... [and] ... a primitive accounting system .... [W]e have lots and lots and lots of different kinds of computers. They have a hard time talking to each other. We have a large number of minicomputers out there that are different kinds, that have different software .... Our system is a road map from hell.

As another example (Stevens 1998), Owens Corning moved to a standard client server architecture – using Internet protocols and off-the-shelf package software – in order to generate a lower cost of system and network ownership. This business case rationale was used by Owens Corning in conjunction with other rationales, as noted below.

Poor-Quality Existing Systems

In some companies, ERP was chosen because the existing systems were of such poor quality that an integrated ERP system offered the only possible opportunity for improvement. Existing systems may have failed or been near failure, forcing the firm to investigate other options.

Case Studies. The apparent inability of Microsoft's systems to meet the demands placed on them led to Microsoft's ERP decision. For example, as noted by John Connors, Microsoft's corporate controller when Microsoft decided it needed to move to SAP:

> We had just had a very bad budget process. ITG [the information technology group] and finance had developed a new budget tool and it didn't work .... It was not fun .... I was just back from vacation, and Steve Ballmer was just back from Wal-Mart. Steve knocked and opened my door. I knew it was Steve, he has a really distinctive knock. He walked in and said, "You guys [expletive deleted]!" I got the message. (Bashein, Markus, and Finley 1997)

Cisco's adoption of Oracle's ERP system (see Cotteleer, Austin, and Nolan 1998) also was motivated by system failure.

> [I]n January of 1994, Cisco's legacy environment failed so dramatically that the shortcomings of the existing systems could be ignored no longer. An unauthorized method for accessing the core application database – a workaround that was itself motivated by the inability of the system to perform – malfunctioned, corrupting Cisco's central database. As a result, the company was largely shut down for two days. Cisco's struggle to recover from this major shutdown brought home the fact that the company's systems were on the brink of total failure.

Difficult-to-Integrate Acquisitions

Acquisition of disparate companies can put substantial strain on the acquiring firm's information systems. Different systems can make comparison of divisions difficult, which in turn makes resource allocation difficult. In addition, different processes coexisting within these systems can hinder exploitation of economies of scale, and different systems and processes can render communication between divisions problematic. As a result, in some cases firms choose to adopt ERP systems in order to facilitate integration of those acquisitions.

Case Study. As noted by Stephen Utoff, a vice-president of planning for Browning-Ferris,

> [w]e wanted more insight into how our processes were doing .... Your processes have to change. As a company that acquired so many companies, [Browning-Ferris] didn't have uniform processes. Part of our challenge was to get 500 places using standard procedures. (Bailey 1999)

Measuring Technical Rationales

Although scaleable, technical rationales generally are measurable on a "yes–no" basis. For example: an application either is Y2K-compliant or it is not; acquisitions either can be integrated or they cannot; systems are either able to support growth or they are not. Technical rationales can provide a strong

Table 7.2. *Anticipated Business Benefits of ERP*

Benefit	Number of Firms	Percent of Sample
Personnel reductions	44	20
Inventory reductions	42	19
IT cost reduction	27	13
Productivity improvements	23	11
Order management cycle time	19	9
Cash management	16	7
Revenue/profit	15	7
Procurement	12	6
Financial cycle close	10	5
Maintenance	8	4

motivation for system replacement, but they provide little guidance regarding system configuration.

Business Process Rationales

Organizations have used a number of business process rationales to justify the choice of an ERP system. For example, an organization might specify a 40% reduction in work-in-process inventory as a potential reason for shifting to an ERP system. As with the earlier survey, Deloitte & Touche also asked firms to specify what business process changes help determine that they needed an ERP system. Business process rationales are largely aimed at improving the organization's efficiency or decreasing costs. The results from that survey are summarized in Table 7.2, where firms could specify more than a single motivation or benefit. In addition to this table, this section provides "drill down" on a number of the business rationales, with some case studies.

Personnel Reductions and IT Cost Reduction

Adoption of ERP systems often is expected to result in personnel reductions, particularly in accounting and information systems.

    Case Study. Geneva Steel's implementation of SAP is expected to have some effects on the number of accountants and information systems personnel. As noted by CEO Joseph Cannon,

> we will have fewer accountants and probably have fewer information systems people. Because one of the things we are considering is contracting out a chunk of that function. A great deal of what we do, we have cost accountants do, lots of

things, not just by hand, it is not that primitive, they do a lot of work that won't need to be done once SAP is implemented.

As a result, Geneva expected staffing of the information technology (IT) department to drop from roughly 80 to 12 and of their accounting function from 60 to 10.

Productivity Improvements

Schaff (1997) remarked that, because improving productivity will always be in vogue, ERP vendors will see continued demand for their products. Accordingly, it is not surprising that some firms' rationales for deciding to go with an ERP system are based on improving particular aspects of productivity.

Case Study. Owens Corning based their adoption of ERP on improving productivity. As noted by Michael Radcliff (vice-president of world wide services and CIO at Owens Corning) when their ERP project began, "[t]o get the project [cost] justified we intentionally focused on the tangible items the board would understand and that we could clearly articulate and make commitments to deliver" (see Stevens 1998). Measurable benefits of up to $50 million were expected based on the following benefits, among others:

- a 1-percentage-point cost reduction deriving from global economies of scale in raw material purchases;
- a 1-percentage-point cost reduction deriving from fewer warehouses and lower freight cost; and
- improvement in reliability-oriented maintenance, yielding lower plant maintenance costs.

Financial Cycle Close

In order to make quality decisions, timely information is necessary. One of the primary measures in this vein is the amount of time that it takes to close the financial cycle. As a result, speeding up the financial close time can be an important motivation for choice of an ERP system. For example, firm XYZ (see Appendix 12-1) wanted ERP implementation to cut their monthly closing process from 24 days to 6 days.

Measuring Business Process Rationales

Unlike technical rationales, business process rationales can employ specific measurement goals that can be used to predict whether ERP will be cost-beneficial or not. In addition, specific process rationales can help in the design and evaluation processes discussed later in this chapter and also in Chapter 12.

The predictability of measurements probably will depend on whether technology enabled or clean slate reengineering is used. If clean slate reengineering is used then it is likely to be very difficult to get any kind of benchmark for expectations. However, if ERP technology enabled reengineering is used then the experience of other firms can be used to guide expectations.

Strategic Rationales

Enterprise resource planning systems can provide the ability to implement different strategies that existing software does not support. These strategies move beyond the efficiency focus of particular business process improvements. Strategic rationales are likely to be based on goals of improving customer response or overall quality, issues not normally associated with transaction processing systems. Further, strategic rationales may lead to implementation of ERP as an information backbone that can be used to provide a base for electronic commerce. This last set of extensions to ERP is discussed in Chapter 14.

Case Studies. Quantum chose Oracle Applications in order to improve their customer response capability. The decision to go with ERP was based on its ability to allow them to execute that customer-oriented strategy and specific ERP capabilities.

As another example, Cisco also apparently views IT expenditures strategically. As noted by chairman John Morgridge, "[o]riginally, we did not think of this [IT] capability in dollar terms. We thought it was necessary to provide quality service, and this is what drove the program" (Stevens 1998).

Competitive Rationale

Although the technology and business process rationales are important, one of the primary reasons for the movement toward ERP is that the competition has it. As noted by Bruce Richardson of AMR Research, "a lot of ERP purchases are premised on the need to just stay in business."

If the competition has ERP and you don't, then the competition may (for example) have superior customer response abilities or may be able to make information available to its managers faster, particularly if they choose to implement in order to accomplish such a specific strategy. As a result, one firm's adoption of ERP can make it necessary for directly competing firms to adopt ERP.

Statement of "the competition has it" as a rationale can take at least two completely different approaches. On the one hand, the firm can simply state that their primary competition has it and thus their firm must pursue it. Alternatively, the firm can focus on why it is important that the competition has it and can examine what specific benefits can be obtained by pursuing a similar option.

Case Study. Suppose that your firm was a competitor of Quantum, which decided to go with ERP in order to improve their customer response. In particular,

ERP allowed them to provide their clients with the "available to promise" (ATP), a unique capability that allowed them to ensure that inventory could be allocated to particular customers – generally, their larger and better customers. Since Quantum was an early adopter, as a competitor you might sit back and wait to see if the software was implemented successfully and if the ATP worked as expected. When the software came on line, you might think about implementing it yourself in order to provide customers the same capabilities.

Western Digital is one of Quantum's primary competitors. On June 24, 1996, Oracle's Application division announced that "[s]everal companies went live with their Oracle Applications implementations during the quarter, including Silicon Graphics, Inc. and Quantum Corporation, both of whom successfully deployed large-scale implementations." At the same time, Oracle's Application division announced that "the customers added this quarter included ... Western Digital ...."

Using Business Case Rationale to Guide Design and Evaluate Success

In some situations, the business case rationale can be used as a basis to guide decisions. In addition, the business case can be used to facilitate evaluation of the success of the implementation; this latter issue is addressed in Chapter 10.

The rationale that was used to guide the choice of an ERP can also be used to guide the design. If the choice of ERP was made based on a business process efficiency or strategic approach, then it is possible to gather information for the design based on the choice rationale. If ERP is being implemented to facilitate a particular strategy, then processes can be designed to accomplish that strategy. For example, if the rationale is to improve logistics then processes should be designed specifically to accomplish that goal.

However, if the choice of ERP is based on technology rationales (such as the need to have software for client server environments or Y2K) or on competitive rationales (say, in order to stay in business), then there is little that can be drawn from the choice of ERP rationale to facilitate design. There are no explicit design choice criteria associated with those ERP choice rationales. As a result, the strategy- and productivity-based rationales are more robust in terms of providing input into the design process.

A Net Present Value Approach

Although each category is treated separately, a detailed net present value approach could be used – as with any investment – to analyze all the costs and benefits involved. Wagle (1998, p. 132) suggests the following five steps.

(1) Create a base case of year-by-year savings from cost cuts that could be made without the ERP system in place.

(2) Create an ERP case of year-by-year savings that could be made with ERP. This should include savings that do not depend on ERP (the base case as in step 1) as well as those that do.

(3) Subtract the base-case savings (step 1) from the ERP case savings (step 2) on a year-by-year basis, and calculate the net present value (NPV) of the residual cash flow. A positive NPV will indicate that you should probably proceed with the deployment of the ERP.

(4) If step 3 produces a positive NPV, conduct sensitivity analysis to ensure that the business case is strong enough to withstand slippage and cost overruns.

(5) Back-allocate all ERP system deployment costs to individual business units so that they can factor them into their planning. Ensure that each unit is held responsible for producing the promised changes.

Rationale Measurement: Monetary versus Nonmonetary Goals

Independent of the business rationale for going with ERP, many firms require some kind of internal (monetary or nonmonetary) measure of the costs and benefits of the ERP system.

Monetary-Based Goals

Apparently, upper management and boards of directors are likely to prefer ERP evaluation in terms of monetary goals. As noted by Michael Radcliff when Owens Corning began their ERP project (see Stevens 1998),

> [ERP is] ... going to cost a lot of money and require a visit from the board of directors. We had made it this far on common sense, strategic insight, collaboration, intuition, and vision, but eventually we had to get down to dollars and cents.

Nonmonetary-Based Goals

Because of their focus on strategies first, some firms (such as Cisco) appear to use a nonmonetary approach when evaluating the potential investment in ERP systems. For example, as noted by Peter Solvik (CIO of Cisco):

> Companies that justify IT projects up front based on ROI [return on investment] are probably not using it strategically .... We use measurable goals, but they are not based on dollars – things like customer satisfaction, reduction in time to close books, getting reports to management faster, or committing to a customer shipment faster. (Stevens 1998)

How Does a Firm Ultimately Decide Whether to Go ERP?

Many of the case examples discussed in this chapter (e.g., Litton Data Systems and Owens Corning) used a cost–benefit analysis to make the ERP decision. However, rather than being a decision-making approach, cost–benefit analysis may prove to be simply a "decision explanation" approach. For example, some consultants have argued that measuring the return associated with different rationales and ERP systems is, in general, not feasible – whether in monetary or nonmonetary terms. As noted by Bruce Richardson, a vice-president at AMR research, "[n]o one has been able to demonstrate any payback from ERP."

Such a concern might derive from other issues. For example, that statement could be a function of the quality of the measurement data (e.g., hard vs. soft data). Alternatively, it might also be a function of how the measurement is used. A third possibility is that the culture of the organization requires a certain approach.

Hard versus Soft Data

The "quality" of the data underlying the rationale can be affected by the type of rationale generating the evaluation data. For example, as noted by Barry Seid of Litton Data Systems, "I had people do advanced studies on return on investment [on various aspects for improving productivity],... but it was soft data, not hard data."

In addition, the quality of the data can influence the potential for generating corporate buy-in. As further noted by Seid, "[t]he Y2K data was hard data. They [top management] believed that data."

Use of the Measurement

As seen at Litton Data Systems, cost and benefit measurement was based on the current ERP project. However, in some cases, returns are used to guide other project investments. John Morgridge (chairman of Cisco) noted that "[g]enerally, we do not do conventional return-on-investment analysis prior to making the investment and launching an IT program. We study the return after the investment and use that as a guide in terms of direction and the amount of future investment in IT" (Stevens 1998).

In any case, measurement of project costs is necessary to evaluate the overall success of the implementation and provide an "actual" for the initial budget. Those project expenses should start with the costs incurred with the "ERP, no ERP" decision and continue through the project life cycle.

Organization Culture

The organization culture may call for detailed a priori financial and nonfinancial analyses of the project. Alternatively, there may be a focus on a posteri analysis, which in turn can be used on future technology efforts. Unfortunately, estimating benefits can be fuzzy and unanticipated; costs can be disguised or hidden. Detailed analyses can therefore result in inaccurate or unreliable data. Throughout, organizations must guard against approaches that undermine what is trying to be accomplished.

The Role of Top Management

What is the role of top management in making the decision to go ERP? First, it is important to realize that only a few executives in an organization can make a decision that is going to cost, on average, $15 million. Typically, such a decision ultimately is made by the CEO, CFO, chief production officer, or board of directors. Second, since there are major changes to domain processes concomitant with implementation of an ERP, the voice of change must come from the domain areas (e.g., production or finance). Thus, top management of those domain areas or the CEO or board must be involved in the actual decision. Third, as noted by a partner in charge of the SAP practice at Andersen Consulting (see Baatz 1996), "[a] SAP project will change lots of people's jobs, and the IT executive alone cannot effect this change."

As a result, it is probably not surprising that, as noted by an industry analyst at Industry Directions, "[b]asically SAP has bypassed the IS department. Instead of doing a technology sell, SAP has gone to the economic buyers [CEOs and CFOs] and convinced them that this software was going to change their business" (Baatz 1996). SAP originated this approach, and other ERP vendors rapidly followed in what has been called "marketing genius" (Baatz 1996).

Summary

This chapter has investigated some of the rationales that have been used to make the decision to do ERP, including

- technology rationales (year 2000 concerns),
- competitive rationales (to stay in business),
- business process rationales (efficiency and productivity issues), and
- strategic rationales (customer service or quality).

The first two sets of rationales provide only limited guidance in process choice required for system implementation. However, if the latter two rationales are used they also can be used to guide design and evaluate success, since they provide greater process specificity and a benchmark to measure improvement against. Ultimately, to gain corporate acceptance, rationales are evaluated using both monetary and nonmonetary measures. Although firms do use these measures, there has been some concern regarding the quality of the underlying data and when to use them. In any case, evaluating the success of a project requires gathering cost data throughout the project life cycle.

References

Baatz, E. (1996). "Marketing Genius." *CIO Magazine,* June 15.

Bailey, J. (1999). "Trash Haulers Are Taking Fancy Software to the Dump." *Wall Street Journal,* June 9.

Bashein, B., Markus, L., and Finley, J. (1997). *Safety Nets: Secrets of Effective Information Technology Controls.* Morristown, NJ: Financial Executives Research Foundation.

Cotteleer, M., Austin, R., and Nolan, R. (1998). "Cisco Systems, Inc.: Implementing ERP." Report no. 9-699-022, Harvard Business School, Cambridge, MA.

Schaff, W. (1997). "PeopleSoft Takes a Hit, But Watch for Comeback." *Information Week,* April 28.

Stevens, T. (1998). "Proof Positive." *Industry Week,* August 17.

Vaughn, J. (1996). "Enterprise Applications." *Software Magazine,* May, pp. 67–70.

Wagle, D. (1998). "The Case for ERP Systems." *McKinsey Quarterly,* no. 2, pp. 130–8.

Appendix 7-1

ERP Choice – In-House or Outsourced

When thinking about going with ERP, perhaps the most basic decision is whether or not to outsource the software. There are two primary ways that ERP can be outsourced. First, the firm may do an analysis and decide that they would prefer to outsource their ERP system, rather than have the design, choice, implementation and support for the system in-house. Second, the entire data processing department may be outsourced, where the decision to implement ERP is made in conjunction with the firm to whom the IT (information technology) department has been outsourced.

Outsource ERP?

When choosing ERP, one crucial issue is "should it be done in-house or should it be outsourced?" Oracle (Stein 1998a), PeopleSoft (Stein 1998a), J.D. Edwards (Stein 1998b), and SAP (Sweat and Stein 1998) have all developed initiatives using their ERP software as an outsourced component.

The outsourcing involves coupling with a partner in order to bring the product to the customers. For example, SAP America is partnering with Andersen Consulting, AT&T Customer Care, and EDS (Sweat and Stein 1998). J.D. Edwards is partnering with World Technology Services, IBM Global Services, and others.

The potential advantage of outsourcing is clear. As noted by a $60-million general contractor (Stein 1998b) that chose J.D. Edwards: "We needed a new accounting package, but we were unwilling to make a huge investment in software and hardware. With outsourcing we avoided a huge financial commitment and minimized our need for MIS people."

However, there are some disadvantages. For example, as noted by a vice-president of IT at Fujitsu Microelectronics, "[t]he key issue is matching up our software to the business and making sure the applications are optimized for our company. I am not sure that a third party outfit could do that for you" (Stein 1998a).

Outsource the Data Processing Department

Generally, the issues of when should a firm outsource their data processing are outside the scope of this book. However, it is of interest to us when the outsourced data processing department decides to implement an ERP system. For example, Herrera (1999) reports that IBM – who took over Kodak's data processing department – will have such a decision to make.

Questions

1. What are the advantages and disadvantages for outsourcing ERP?
2. To what kind of firms is outsourcing ERP likely to appeal?
3. What is likely to drive ERP adoption for those firms with outsourced IT, such as Kodak?

References

Herrera, S. (1999). "Paradise Lost." *Forbes Global,* February 8.
Stein, T. (1998a). "PeopleSoft Ventures into the Unknown World of Outsourcing." *Info-World,* January 19, p. 35.
Stein, T. (1998b). "J.D. Edwards Jumps into Outsourcing." *Information Week Daily,* February 12.
Sweat, J., and Stein, T. (1998). "SAP Formally Rolled Out a Program ...." *Information Week Daily,* February 24.

8

Choosing an ERP System

This chapter examines how firms choose between different ERP systems. Two primary approaches are used to guide ERP choice: requirements analysis and gap analysis. This chapter defines these approaches and examines the advantages and disadvantages of each, and it briefly discusses an alternative approach. In addition, this chapter examines the assumptions behind requirements analysis and gap analysis as well as some limitations of these two processes. Also discussed are two companies that have mitigated these limitations in their evaluation of ERP software by going beyond requirements and gap analyses. Finally, this chapter summarizes the shortcomings of broader evaluation approaches.

Requirements Analysis

Requirements analysis is a review of system requirements for organizational models, artifacts, and processes (MAPs). In some cases requirements are cataloged as to their importance – for example, required or optional, or ranked (say) from 1 to 3. Requirements are then summarized in a requirements document (request for proposal, or RFP) that is provided to different vendors. The organization uses that set of requirements to judge how well different pieces of software meet their needs.

How Many Requirements?

Requirements documents can be quite extensive. For example, one company (Firm A)[1] with $40 million in sales produced a document with about 1,000 requirements listed. Timberjack,[2] with over $35 million in sales, listed 1,042

---

[1] This chapter refers to some firms that would rather not be identified by name. As a result, they have been labeled Firms A, B, and C.

[2] This section and other sections in this chapter make reference to "Timberjack," which is discussed in Romanow, Keil, and McFarlen (1998).

requirements. A much larger and privately held company (Firm B) also had a requirements document with over 1,000 items. It is likely that the number of requirements is a function of firm size, specific industry standards, and who is doing the analysis.

Time to Develop Requirements

Because of the large number of requirements and the number of processes covered by an ERP, developing a requirements document can require substantial time and cost. Firm A was able to do its complete requirements analysis in a single month. Timberjack's requirements analysis took roughly three months. Firm B took four months to develop their requirements document.

Who Should (or Does) Develop Requirements?

Who develops the requirements? Typically, requirements are generated by someone very familiar with the existing process, and often someone who is or will be doing the work encompassed by the particular set of requirements. In some cases, process managers are brought in to generate models of existing processes.

Who should develop the requirements depends on whose perception the organization is most interested in. If day-to-day personnel are used then there is likely to be a set of requirements that mirrors existing processes. If management is responsible then the requirements will likely not mirror actuality as closely, since management is withdrawn from day-to-day operations. Management may generate requirements that it thinks "should be there" or requirements that "used to be there" when the manager worked in the process on a day-to-day basis.

At Firms A and B, the day-to-day process users worked in conjunction with a consultant to gather the requirements. At Timberjack, the operations were being moved to a different city. As part of this project, a consultant worked with a group of day-to-day process users who were not going to be using the new system. At Timberjack, there apparently was concern for the potential loss of knowledge of their day-to-day processes.

Requirements analysis often employs consultants – both for their general knowledge of the requirements analysis processes and for their "arms-length" views into the process. Perhaps the best team is one that mixes both management views and current day-to-day process worker views with a consultant who provides experience and an independent vantage point.

Help for Doing Requirements Analysis

In addition to consultants, software is available that tracks different requirements in competing vendors' software. For example, The Requirements Analyst

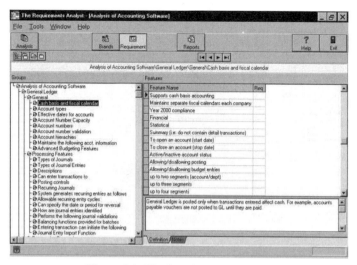

Figure 8.1. The Requirements Analyst

tracks a wide range of capabilities of different accounting packages and modules within those packages. Users can choose which capabilities are required and then generate comparisons between their requirements and a vendor's software capabilities. A sample screen is shown in Figure 8.1 (www.ctsguides.com).

Requirements Granularity (Fineness)

Requirements vary substantially in their granularity, also referred to as "fineness." For example, in some cases a requirement may reference a set of best practices, whereas in other cases a single data element can be specified. For example, Timberjack requirements included the following:

- able to manage orders following best practice methods of placement, control, and expediting;
- able to use EDI (electronic data interchange) with certain vendors;
- able to set up vendor schedules;
- able to track actual date.

Advantages of Requirements Analysis

There are a number of advantages of requirements analysis. First, it results in a list of requirements that can be used to benchmark different software packages. Using the requirements analysis, software can be examined to see what is gained or lost when compared to legacy or alternative ERP software. Second, requirements analysis provides a basis for communication and discussion of different MAPs. Accordingly, it can facilitate inter- and intradepartment communication. Third, requirements analysis often results in a better understanding of the limitations of existing MAPs; this, in turn, can prompt increased awareness of a need for change. Fourth, requirements analysis can be used to attain buy-in to the new system by ensuring that specific needs are met. When users participate in the process they have input to the process and – it is hoped – more confidence in the resulting system choice. Fifth, since the organization is currently using the "discovered" requirements, the functioning organization provides evidence that the MAPs do work individually and as a portfolio.

Disadvantages of Requirements Analysis

There are also some disadvantages of requirements analysis. First, requirements analysis can be very time-consuming, delaying the choice of a system. In the examples just discussed, it took from one to four months to develop the requirements document. Second, requirements analysis can be very costly. It is not unusual for such engagements to result in consulting costs of $100,000 or more, and that does not include the cost of time of contributors from the firm for which the requirements are being written. Third, if the requirements document is too large or has too many requirements, ERP vendors may see themselves as too busy to respond. In particular, the vendors with the most time (i.e., with the most unused people) may be the only ones choosing to respond. If the requirements document is too large, firms may be limiting the set of potential respondents. Fourth, requirements analysis may end up specifying requirements that lock the firm into dated ways of doing things. Instead, it might be more beneficial for the firm to reengineer existing processes. Fifth, processes (and their interfaces) change, so this type of analysis is likely to be chasing requirements that are never totally stable. Requirements analysis may provide at best a "snapshot" of a firm's processes.

Format of Requirements

Requirements analysis takes at least two different forms. Perhaps the most frequently used approach is to prepare a detailed list of all that is required from the software. That list is then sent to the vendors so that they may compare it with their software's capabilities.

Another approach that is gathering increasing momentum is the "scripted" approach, which generally has one and sometimes two phases. The first phase is loosely scripted; in this phase, relatively fixed ERP demos (developed by the vendors) are shown to the client. The client has very little control over the demo but can ask for demonstration of specific features of particular importance to the firm.

Typically there is a second, more tightly scripted phase. In this demonstration, the ERP vendor is asked to use specific data and perform particular activities. For example, if there is interest in project monitoring capabilities then the firm would ask the vendor to set up a project in the software. This approach is particularly useful for identifying gaps that might necessitate modification.

Gap Analysis: Comparing "As Is" and "To Be"

An alternative approach is to develop an "as is" analysis and a "to be" analysis and then compare these two via a *gap analysis.* "As is" refers to current system functionality. The firm must then plot a course from the "as is" to the "to be" model. The "as is" model provides a benchmark for current system capabilities. The "to be" model can be either clean slate or technology enabled. If the "to be" model is developed independently of a specific package, then it is clean slate. If it is a clean slate approach then the "to be" model becomes equivalent to the requirements model. Increasingly, however, "to be" analysis is based on a specific ERP package; in this case, "to be" analysis is technology enabled. Hence, when based on packaged ERP software, the "to be" analysis may also be called "best practices" analysis.

Gap analysis is designed to compare the "as is" and the "to be" analyses to determine what gaps exist between what the company now has and what they've decided they need. Ideally, clients can use this information in choosing software that best meets their needs. As mentioned previously, identifying gaps enables a firm to determine what modifications might be necessary for software to meet the requirements of the company. In turn, identification of those modifications could be used to better realize the costs – beyond the software – of a specific choice.

However, there are many perspectives on these gaps, and even on what constitutes a gap. For example, given a list showing what the firm needs and what the vendors offer, how should we proceed? Should we count the requirements and see who meets the most, with the winner the software choice? Given the lack of standard granularity, can we even "count" at all? Alternatively, should the focus be on only the most critical requirements? Part of the process of doing gap analysis is that these questions are addressed by the specific firm making the choice. However, some are beginning to ask: Should there even be a gap analysis? Or should another approach be used?

Emerging Approach

Increasingly, consultants are urging companies to not do an "as is" analysis. Instead, ERP implementation is seen as true technology enabled reengineering that goes straight to the "to be" process using a specific ERP package. The argument goes as follows.

(1) Each of a set of ERP packages is assumed to meet organizational needs. Selected ERP software is assumed to have roughly the same set of best practices and be roughly equivalent.

(2) Replicating existing processes is of little use. Organizations should not replicate existing, dated processes but instead should reengineer using the capabilities of the ERP system. Accordingly, organizations should see the implementation of an ERP system as a chance to improve their existing processes in the light of new technology.

(3) A priori best practices analysis has no place in the implementation of ERP systems. Instead, best practices analysis should take place only within the context of the specific piece of ERP software chosen.

(4) Since implementation costs are minimized by making few changes in the software, only best practices available within the specific ERP system should be chosen.

How Does One Actually Choose Software?

Still at issue is whether the software will actually meet organizational needs. As a result, in this approach additional information is employed in the choice process. First, are there any similar firms (or divisions within the firm) that have adopted an ERP system? If so, what system did they adopt and how well does it meet their perceived needs? Second, what software does the consultant that you trust recommend? In some cases, the few consultants that you trust may have detailed expertise in a particular ERP system, which they are likely to recommend. (Not surprisingly, a number of consultants have told me that this is the best approach.) Third, if each of the ERP packages has virtually the same capabilities, then this may not be a bad strategy – and likewise for a firm with relatively generic needs.

Regarding some processes, however, different ERP software packages have different features (e.g., "available to promise" as in the Quantum case). In such settings, finding the software with unique features could become a critically important factor. On the other hand, as time goes by, there is likely to be a convergence in ERP software capability, with fewer and fewer differences between the various vendors' packages.

Problems with Requirements Analysis and Gap Analysis

Although requirements analysis and gap analysis have a number of advantages, both make some important implicit assumptions and ignore a number of important issues.

Assumptions Implied by Both

Both forms of analysis make some critical assumptions that should be examined by any firm employing them. First, both approaches generally assume that the preferred ERP package is the one with more of the listed features. Unfortunately, this ignores other viewpoints: perhaps the focus should not be on all characteristics but instead on the module (or modules) that are the most important to value creation.

Second, granularity of requirements is assumed away, particularly when organizations tote up the requirements satisfied. As a result, in counts comparing how many requirements are met, a data element might be equated with an entire process. Bottom line: simply counting requirements met does not provide sufficient evidence for choice of a software package.

Issues Ignored by Both

Requirements analysis and gap analysis generally focus on functionality. As a result, they typically ignore a wide range of other factors, including the following.

Cost. Analysis seldom includes cost. Typically, analysis points to a particular vendor or multiple vendors, and only then is cost feasibility assessed. Analysis efforts often appear to be done without regard to the software cost. However, one of the most critical relationships in virtually every business setting is that of cost/benefit. Ignoring costs may be a function of who is doing the evaluation. If evaluation is being done at lower levels in the organization but the budget is controlled in the upper levels, there may be insufficient capability of bringing the two together.

Installation Time. In some situations, organizations may face time constraints due to their desire to catch the competition or be a first mover. Oftentimes, issues such as installation time are not addressed until after the software choice.

Flexibility. When used in the context of ERP, flexibility generally refers to the ability to use numerous best practices. Typically, the more best practices available, the more flexible the software.

User Interface. The user interface comprises both inputs and outputs. Naturally, firms prefer easier-to-use systems. As discussed earlier, ERP vendors have increasingly tried to increase their systems' "user-friendliness."

Upgradability. There are at least two issues that relate to upgradability: How often do new versions become available, and how easy is it to upgrade to new versions? Upgrades that are either too frequent or too infrequent can cause problems. If updates are too frequent, then firms are constantly faced with up-grading their system to obtain new capabilities and wondering how long their earlier versions will be supported. If the updates are too infrequent, firms may not get the functionality that they need.

Computing Environment. Computing environment or philosophy can be an important issue. Reportedly, Oracle's focus on a "thin client" computing environment had a negative impact on some experts' opinions of the product. Computing environment may not be an issue for all firms, but in some cases ERP systems will need to interface with other systems. If so, then interface capability is likely to be a critically important differentiating factor that needs to be evaluated.

Implementation Personnel. In most ERP projects, consultants play a critical role. As a result, the quality and availability of those personnel can influence the choice of the software. Thus, some firms include implementation personnel as an important evaluation variable.

Day-to-Day Use and Maintenance. After an ERP system has been implemented, there is a shift to maintenance and day-to-day activity. The "care and feeding" of particular ERP systems may differ. As a result, this set of factors should also be included in the analysis. Unfortunately, the choice group may differ from the implementation group, which in turn is likely to differ from the maintenance group; hence, the significance of this factor is variously interpreted.

Functionality. As will be seen in the next section, some firms may try to measure functionality as a single characteristic and not on a requirement-by-requirement basis. When this is done, there may be miscommunication as to what constitutes "functionality," which is likely to be in the eye of the beholder.

How Do Organizations Evaluate and Choose ERP Software?

The effort by firms to include a broader base of factors (beyond functional requirements and gap analysis) will be referred to as "evaluation." Since requirements analysis and gap analysis focus on function, and since there clearly are a range of other issues, how do organizations evaluate and choose ERP software? This section presents the approach used by Timberjack (Romanow et al. 1998); another case using a different method is given in the appendix.

Timberjack

Timberjack used a set of evaluation criteria and weights as part of their choice between QAD and Oracle Applications. A small group (consisting of four employees of Timberjack and a consultant from Coopers & Lybrand) was chosen

Table 8.1. *Evaluation Criteria and Weights*

Factor	Weight	QAD	Oracle
Support	20		
Functionality	30		
User interface	10		
Flexibility	20		
Future prospects	10		
Reliability	20		
Integration	30		
Platform	20		
TOTAL	160		

to develop an approach for choosing between QAD and Oracle. The group decided on a set of factors and weights that were used to generate comparisons between QAD and Oracle, with the decision going to the system with the largest total weight. The factors and weights are summarized in Table 8.1.

Evaluation Issues to Be Considered

Evaluation beyond requirements analysis and best practice analysis involves a number of issues that must be carefully considered lest the process be manipulated.

Using a Single Number to Evaluate Packages

The advantage of this approach is that it provides a single number that can simply be attached to each ERP system being evaluated. However, it is arguable how sensible it is to attach a number to such aspects as "future prospects" and other factors. In addition, it may not be sensible to combine numbers across such diverse categories and expect their sum to represent a valid ranking of the ERP system.

Manipulation of Factors and Weights

Unfortunately, the factors that are chosen and their corresponding weights can be manipulated. Given information about two or more ERP packages, the factors and weights can be manipulated in favor of a particular package or manipulated away from other packages. For example, Timberjack assigned only 30 (out of 160) points to functionality. Such a low weighting does not benefit systems with high functionality – the primary focus of requirements and gap analyses. This shows how the factors used in the evaluation process can be manipulated so that a particular choice is ensured. Moreover, the scope of the set of factors considered may be too large or may not be complete. For example, Timberjack chose eight factors, which did not include implementation personnel; their list excluded other factors as well, such as maintenance.

Where Does the Variation Occur?

In order to facilitate choice between alternative systems, variation must be found. As a result, firms may find it advantageous to focus on those issues that differentiate one package from another. Is there a single feature that can drive the choice process? This was seen with Quantum in their choice of Oracle Applications because of its "available to promise" capability. After these differences are found, the firm may then determine which (if any) are important.

Choice versus Explanation

Factors and their weights are a tool for either helping to choose software or explaining the choice, depending in part on when the factors were chosen and the weights assigned. If the factors and weights are chosen before the software is evaluated then they provide a basis for choice, based on an a priori understanding of ERP software and the organization. If the factors and weights are chosen after the software is evaluated, they may provide a way of explaining or rationalizing choice.

Who Should Generate the Factors and Weights?

A consultant or group of consultants is often added to the selection project. The roles of the consultant may include providing independence and ERP system knowledge to the process. However, the consultant is also a variable that can be manipulated. For example, particular consultants may be chosen owing to the consistency of their world view with that of the choosers. Moreover, those who vote or set the weights can themselves be manipulated or politicked. Adding people with specific viewpoints can push ERP system choice in one direction or another.

When Should Factors and Weights Be Generated?

In some settings, the importance of some factors may not be determined until the capabilities of ERP systems in general are understood. Furthermore, it is only by examining system capabilities and reviewing organizational preferences that such lists can be constructed. Lists developed prior to organizational and ERP system examination are not likely to reflect the depth of understanding available after their examination.

Table 8.2. *ERP Costs*

	Upgrade Existing	Oracle	SAP
Implementation	$3–5 million	$4–8 million	$6–10 million
Specific software development fit	$2–4 million	$1–3 million	minimal
Number of people to maintain	15–20	10–15	8–10

Cost Factors

At a budgetary level, the most important factor is cost of the software. Even so, as noted before, in some cases that cost is ignored. How do firms take cost into account as part of the choice process? One approach (used by Firm B) was to list three types of costs and use their total to facilitate the decision (see Table 8.2). (Firm B ultimately chose SAP.)

Although this approach considers economic factors, it ignores benefits. For example, it assumes that – after development for specific fit – each piece of software will provide roughly equivalent benefit. Further, it assumes away the potential benefit associated with changing business processes via ERP software. It also ignores the costs to educate users on how to use a new system (and, conversely, the benefits of *not* having to re-educate their users by staying with the old system).

What Is Unique to ERP?

The requirements approach discussed in this chapter is hardly unique to ERP. However, ERP does provide a wide range of best practices, and choosing software for those best practices is relatively unique for ERP systems. Now that ERP systems have originated this best practices approach, other forms of software are pursuing a similar strategy. In addition, the broad base of best practices is putting ERP systems in a unique category of software – where some consultants feel comfortable recommending the software even without a requirements analysis.

Summary

This chapter has investigated how firms choose ERP systems. Typically, firms will do at least one of a requirements analysis or a gap analysis. However, both have not only advantages but also disadvantages. For example, requirements analysis can replicate dated processes that should be reengineered. Gap "to be" analysis, unless structured within a specific ERP, may generate an infeasible portfolio of best practices. Enterprise resource planning software has an increasingly broad range of capabilities. As a result, an alternative strategy is simply to choose one of the better ERP packages that a consultant knows and has already implemented elsewhere.

Requirements analysis and gap analysis focus primarily on functional characteristics; in so doing, they ignore a broad range of other factors. This chapter has discussed some of these overlooked issues and has examined the approaches used by two firms in their choice of ERP software.

Reference

Romanow, D., Keil, M., and McFarlen, W. (1998). "Timberjack Parts: Packaged Software Selection Project." Report no. 9-398-085, Harvard Business School, Cambridge, MA.

Appendix 8-1

Chesapeake Display & Packaging

Chesapeake Display & Packaging (CDP) is a subsidiary of Chesapeake Corporation. Gary Cheimis, vice-president and chief information officer for CDP, set up a five-step process for enterprise software selection:

- form a small, blue-ribbon team;
- contact vendors to arrange demos;
- ask the vendor for proof of rapid implementation capability;
- vote;
- make the choice.

Form a Small, Blue-Ribbon Team. A blue-ribbon team (BRT) was formed to choose the software. People were chosen for their knowledge of the business and business processes. People carried a mix of views; some were big-picture people and some small-picture people. The BRT was limited to no more than ten individuals.

Contact Vendors to Arrange Demos. A limited number of first-tier enterprise software vendors were chosen, contacted, and asked to prepare a demo

This appendix is based on G. Cheimis (1999), "Shooting the Rapids of a Rapid Implementation," in *Shape Your World, Focus '99* (Denver, CO: J.D. Edwards).

Table 8.3. *Software Preferences at Chesapeake*

	Division[a]	Position	Functional Fit			Implementation Personnel		
			BAAN	JDE	SSA	BAAN	JDE	SSA
Gary Cheimis	CDP	VP CIO	3	2	1	2	3	1
Elvis Brannam	CDP	IS Mgr	2	3	1	3	2	1
Linda Witter	CDP	Sales Support	3	2	1	3	2	1
Ted Samoits	CDP	Prodtn Mgr	2	3	1	2	3	1
Jack Kirk	CDP	VP CFO	2	3	1	2	3	1
John Polgar	CDP	Sr. Analyst	2	3	1	2	3	1
Carl Wilcox	CDP	Design Adm	3	2	1	3	2	1
Richard Hastings	CDP	Oper Mgr	3	2	1	3	2	1
Mary Gene Simmions	CC	Dir Shrd Svc	2	3	1	2	3	1
Dick Fuss	CC	Analyst	3	3	1	2	3	1
Randy Grahm	CC	Internal Audit	1	3	2	1	3	2
Bill Tolley	CC	CFO	2	3	1	2	3	1
David Spencer	CPC	Engr Mgr	2	3	1	2	3	1
Ann Walsh	CPC	IS Mgr	2	3	1	2	3	1

[a] CDP, Chesapeake Display & Packaging; CPC, Chesapeake Packaging Corporation; CC, Chesapeake Corporation.

for the BRT. Vendors were given unlimited access to the BRT until the demos were to be conducted, in about three weeks. Demos were set to be conducted over a one- or two-day period. Vendors were allowed to choose the hardware and database system that they preferred.

Ask the Vendor for Proof of Rapid Implementation Capability. Since rapid implementation was required by CDP, vendors were asked to prove their implementation capabilities at the demo. Three such requirements were established:

- show that your software can handle our business;
- show that you can implement in the time required;
- show expertise in understanding the industry.

Vote. After the demos, there were three primary candidates: BAAN, J.D. Edwards (JDE), and SSA. In order to choose between the different enterprise software candidates, the BRT was asked to vote for their preference based on two issues: best functional fit and best implementation personnel. In an effort to keep the voting simple, each person was asked to rank each of the different software options from 3 to 1, where 3 denoted the highest rank. Each person was asked to make the evaluation based on their area of expertise. The votes are summarized in Table 8.3.

Software Recommendation. J.D. Edwards was seen as the consensus choice of the three enterprise systems, since it had the highest total number of points. Qualitatively, Cheimis indicated that the J.D. Edwards package was seen as offering

- superior financial capabilities,
- one integrated solution,
- included human resources and payroll,
- advanced object-oriented tool set,
- Synquest advanced planning module, which allowed for scheduling on a cost basis.

Questions

1. Which ERP system do you think Chesapeake will choose? Why?
2. What are the strengths and weaknesses (if any) of Chesapeake's approach to choosing software?

Appendix 8-2

A CFO's Inquiry

We are at the stage of evaluating different ERP systems. We had a short list of different suppliers of software and in the end we decided to take a deeper look at SAP. Together with a consultant (Coopers & Lybrand), we worked out a three-stage project:

(1) "as is" analysis;
(2) "to be" analysis;
(3) matching process (match the two analyses).

We came to the conclusion that SAP did not meet our demands. In particular, the production module was far below our expectations and did not give us the information that we have now. There are many gaps between the "as is" and "to be" analyses. Further, SAP (owing to its integration) is not as flexible as our system is now.

We have a tremendous amount of information in our current system available in real time. In addition, our current software has been written by our own people and is "consumer oriented." The online information capability of our current system is just fantastic and plays a key factor in our customer relationships. Unfortunately, SAP's production module does not compare.

Since there are a number of unmatched areas, there will be a lot of programming if we want to have what we have now in our current systems. As a result, we asked Coopers & Lybrand to give us a detailed overview of the project cost. If we have to buy SAP only for the financial modules and purchasing modules, the system is too expensive.

The questions I have are as follows.

- Is it worth it to buy an integrated system, where a lot of programming still has to be done?
- Would the project become too costly?

The decision we are going to have to make very soon is certainly a difficult one. Are there any alternatives? Some advice would be welcome.

9

Designing ERP Systems

Should Business Processes or ERP Software Be Changed?

As part of choosing ERP software, firms face a design "reengineering policy" decision regarding the amount of software customization and organizational reengineering that must be done. On one hand, since no software is likely to meet all of a firm's needs, some firms have chosen to substantially customize the software to fit their processes. On the other hand, many businesses have used adoption of new software as a chance to change their basic business processes, reengineering their organization to match the "best practice" processes in the ERP software. In order to clarify the extent to which different firms pursue different policies, Forrester Research conducted a survey (Lamonica 1998) and found the following percentage of firms following policies:

- choose applications that fit their business and customize a bit – 37%;
- customize applications to fit their business – 5%;
- reengineer business to fit application – 41%;
- no policy – 17%.

This chapter analyzes two basic questions.

(1) How does the reengineering policy influence the software application choice process?
(2) Should business processes or ERP software be changed?

How Does Reengineering Policy Influence Software Choice Processes: On the Relevance of "As Is" and "To Be" Modeling

The decision to modify software or reengineer business processes (or both) influences the relevance and use of "as is" and "to be" modeling. As a result, the reengineering decision needs to be coupled to the decision to choose an ERP

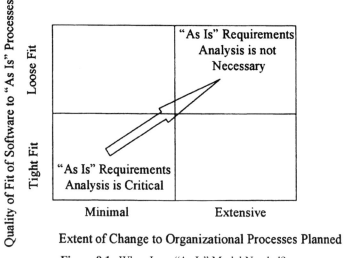

Figure 9.1. When Is an "As Is" Model Needed?

system. Since "as is" analysis generates models of existing processes, the relevance of the "as is" analysis depends on the extent to which the processes will stay the same. Similarly, since "to be" analysis generates models of processes that the organization has chosen to implement (or to replace or supplement existing processes), its relevance depends on the extent of change to be made from existing processes.

On the Relevance of "As Is" Models for ERP Systems

Firms that plan to keep their existing processes – and do little or no reengineering – benefit from a substantial "as is" analysis to make sure that the software meets their needs. In such situations, the tighter the fit to existing processes, the better. However, if a firm is planning extensive reengineering of its processes (as were over 40% of the firms in the Forrester survey), then the "as is" model loses some of its importance. Because many firms that implement ERP plan on reengineering their processes, there is less interest and need for development of a classic "as is" model. As a result, increasingly firms are not performing "as is" analyses. These considerations are illustrated in Figure 9.1.

There are potential problems associated with using an "as is" model. Consider a firm that performs an "as is" analysis and finds a loose fit between existing processes and the ERP software they choose. If minimal reengineering is planned, then the organization may have lost a chance to choose software that matches their processes. Alternatively, consider a firm that does an "as is" analysis and intends extensive reengineering of those existing processes.

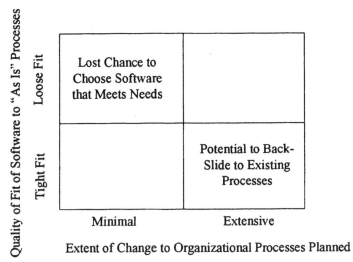

Figure 9.2. Potential Problems of Using an "As Is" Model

Choosing ERP software that has a tight fit with existing processes can limit the ability to do extensive reengineering. If the firm has any trouble implementing new processes then there could be backsliding – away from reengineered processes to a version that employs existing processes. These two settings are summarized in Figure 9.2.

On the Relevance of "To Be" Models

If there is only limited modification of the software planned, then choice of "to be" processes typically is constrained to those that are available in the ERP software. In this limited change setting, "to be" analysis is a choice problem from the portfolio of processes available in the ERP system.

However, if the software is to be substantially modified, then the "to be" process changes from a portfolio selection problem to one of process development. Rather than being technology enabled, the "to be" analysis becomes more like clean slate reengineering with only limited dependence on the processes in the ERP software. This dynamic is summarized in Figure 9.3.

Organizational and Software Change: Small-r versus Big-R Reengineering

Our analysis of the relevance of "as is" and "to be" models yielded two primary dimensions: the extent of change of organizational processes and the extent of change of the software. These two variables, which measure the extent

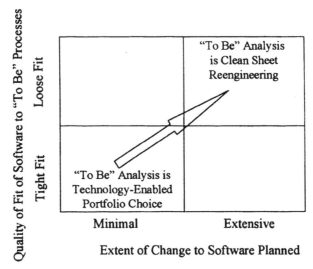

Figure 9.3. What Kind of "To Be" Model is Needed?

of change or reengineering necessary, provide a basis for further comparative investigation.

Aidan Waine, who was the SAP project manager for Microsoft's Windows NT implementation of SAP, referred to that implementation as "small-r reengineering with SAP's R/3 as the change agent." Microsoft implemented SAP's financial modules, using SAP-available processes, with no modification of the software. In addition, because Microsoft implemented only the financial package, there was a fairly tight fit between their processes and the system capabilities. If Microsoft did "small-r" reengineering, what is "big-R" reengineering? Big-R reengineering would be extensive modification of organizational processes and ERP software. With small-r reengineering, the organization adapts itself to tight-fitting software; with big-R reengineering, however, the organization changes processes and software to meet its needs. These two points of view are summarized in Figure 9.4.

The next four sections discuss each of the quadrants of the model depicted in the figure, providing examples, advantages and disadvantages, and characteristics of firms that are likely implementors.

Minimal Organizational and Software Change:
Small-r Reengineering

Small-r reengineering occurs when a firm adopting ERP software makes only minimal changes in that software and minimal changes in the specific organization's business processes when implementing the software. In order to do

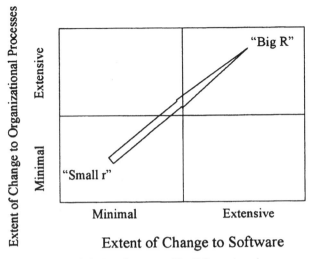

Figure 9.4. Small-r versus Big-R Reengineering

small-r reengineering, the software must be robust enough to adapt to the organization's processes captured in the system, and the organization must be able to adapt to minor changes necessitated when the software does not exactly meet the current organization processes. With small-r reengineering, there is likely to be a high degree of matching between the software and the current organizational processes.

Example

A classic example of small-r reengineering is the Microsoft case. Microsoft implemented SAP's R/3, making as few changes to the software as possible. The implementation view was that the software provided best practices that were acceptable to Microsoft. In some cases, accepting the business practices implied by the software meant changing the organization. However, overall there were apparently few changes made to Microsoft's business processes and organization.

Advantages and Disadvantages

Small-r reengineering offers the advantage of faster and cheaper implementation, because there is no need to change the software or the organization processes. In addition, since other firms are likely to have followed the same or a similar path before, there is experience that can be used to facilitate prediction of implementation time and cost.

There are at least two disadvantages to small-r reengineering. First, since processes don't change, the number of people may not change. As a result, personnel expenditures may not be affected, perhaps contrary to expectations. Second, even small-r reengineering projects that implement ERP can be costly. Hence, small-r reengineering firms may have missed a major chance either to alter existing organizational processes or to adapt the software to meet their unique requirements. Since large expenditures on software are likely to occur only at distant points in time, small-r firms may have missed their window of opportunity.

Likely Processes

Some generic processes are more likely to be small-r rather than big-R. For example, many firms have similar requirements for their financial processes. Processes used for accounts payable vary little from organization to organization prior to any focus on reengineering. As a result, standard and generic processes are likely to be small-r implementations.

Furthermore, processes for which small-r reengineering is used are likely to include those for which there is little value created by using unique processes. Again, financial and "back office" processes are less likely to be seen as a source of value creation and thus will be treated as small-r opportunities.

Likely Implementors

Certain types of firms are more likely to employ small-r reengineering. Implementors of small-r reengineering are likely to be those firms implementing generic (e.g. financial) processes or firms for which information technology is not seen as a critical value-creation process. Best practices built into ERP software generally have been used by other firms. As a result, firms using small-r reengineering are employing best practices that have already been installed by other ERP implementors. Accordingly, those firms using small-r reengineering are not likely to develop processes that provide any unique sustainable competitive advantage. Since the software implemented uses available processes, much of the implementation could be replicated; as a result, any advantage would not be sustainable.

Implementors of small-r reengineering are also likely to be small- to medium-size firms that are "software takers." That is, they basically take software as it is given and do not have enough power or resources to induce software firms to develop special software for them. Implementors of small-r reengineering might also be larger firms that (for whatever reason) do not have the internal patience, resources, or interest in changing either their organization or the software. For example, in the case of Microsoft, there had been two previous projects designed to implement client server ERP systems. Substantial resources and patience had been expended on those previous implementations.

ERP Vendor Interest and Concerns in This Area

Enterprise resource planning vendors such as SAP have expressed great interest in firms implementing tools that facilitate small-r reengineering. For small- and medium-sized firms generally, SAP recommends not making changes in the software. For example, SAP developed the ASAP methodology in order to facilitate such firms' implementations of SAP.

Extensive Organization Change and Minimal Software Change

Some suggest that one of the most important views of ERP software is that the software enables and facilitates business change. For example, as noted by Kevin McKay, president of SAP America:

> SAP R/3 customers often have to change their businesses to use the software, but the cost and the change is worth it because the software lets the company operate more efficiently .... It's not an [information technology] project, it's a business transformation. (Bailey 1999)

Although ERP provides organizations with the "opportunity" to change their business, some firms in this quadrant do not see the need to change processes as an opportunity. Instead, they see it as a constraint on the implementation and limitations of the software. If the software were more fully developed to include a broader range of processes, then many organizational process changes would no longer be necessary.

Examples

The *Wall Street Journal* (Bailey 1999) recently documented the situation of a group of firms in the trash hauler industry that were taking their SAP "software to the dump." The two leading firms in this industry (Allied Waste and Waste Management) were implementing SAP's R/3 but abandoned the initiatives. One of the primary concerns was noted by Thomas Van Weelden, CEO of Allied Waste: "They [SAP] expect you to change your business to go with the way the software works."

This quadrant does not always result in firms "trashing" their SAP efforts. For example, as noted by David Edmondson, assistant provost of information services at Texas Christian University: "We've customized the software very little. Our systems integrator is helping us to change our business processes. We'd rather do that than modify our software" (Lamonica 1998).

Advantages and Disadvantages

There are a number of advantages to changing processes rather than the software. First, not making any changes in the software will facilitate updating ERP software as subsequent versions are released. Second, changing ERP software is a difficult task. Modules are integrated, so a change in one module can require changes in other modules. Third, changing the software means that the firm likely will need to continue maintaining the software changes, adding another layer of expense and requiring expertise that otherwise might not be needed. Fourth, constraining processes to those available in an ERP system can give firms a chance to improve and standardize their processes.

However, there also can be disadvantages to changing processes to conform to the software. First, there is concern that value-creating processes are being changed to generic processes. Reportedly, a number of firms have changed sales compensation from incentive systems to wages in order to have the software fit the process. Unfortunately, such a fundamental change in processes may lead to dysfunctional behavior. If processes help create value then changing them may have a negative impact on the firm. Second, as noted in the previous example, organization process changes might not be successful. In this second case, as seen with the trash haulers, the organization lost the benefit of the use of the software.

Likely and Unlikely Implementors

The likely implementors in this quadrant are those firms that perceive changing people as easier than changing software – or at least as more cost-effective. A PricewaterhouseCoopers consultant specializing in ERP systems indicated that "[i]t is easier to change people than it is software .... Change management becomes a critical tool to facilitate successful process change."

Changing business processes may be difficult in a number of settings. For example, if a firm is heavily decentralized (as were the trash haulers), then changing processes can prove to be problematic because it is difficult to choose (and enforce) common processes for diverse settings with diverse requirements.

Minimal Organization Change and Extensive Software Change

Rather than changing the organizational processes to be in concert with the software, some firms may choose to change the software to match their existing processes or to accommodate other best practices not available in the ERP software.

Examples

A project manager at Nestlé indicated that their choice of processes for their SAP implementations included best practices beyond those included in the software. In particular, some of their existing processes and best practices from their consultants' databases were chosen, forcing a change in the software.

As another example, the manager of SAP services at Deere Company noted that "[w]e did a lot of customization because we have a history of that and we do it pretty well. It's worked well and satisfied business needs" (Lamonica 1998).

Advantages and Disadvantages

Changing the software has a number of drawbacks. Changes make it difficult to maintain and upgrade the software. As noted by Steve Cooper, director of enterprise systems at Corning,

> [w]e have learned the hard way, if you modify the software there will be a cost. The cost comes when you do the modification initially, when you do an upgrade, and when you support the software over time. (Lamonica 1998)

Customization to different divisional requirements also can make it difficult to implement the software in other divisions. Although the manager of SAP services at Deere Company indicated that the customizations went well for their ERP project, it was also noted that "what hasn't worked well is establishing standards and templates that can be rolled out to other divisions" (Lamonica 1998).

Likely Processes and Implementors

Value-creating processes that differ from company to company are likely to be the source of changes in the software. For example, production management modules are likely to be changed or links to legacy systems are likely to be made.

Company size has a major influence on the firm's ability to change either the software or its own organization. Larger companies have greater resources, so they can more easily focus on software changes. Further, as noted by the director of information technology at Penwest Pharmaceuticals, "[b]ig companies have less flexibility to change their company to fit the software" (Lamonica 1998).

Extensive Organization Change and Extensive Software Change: Big-R Reengineering

Big-R reengineering occurs when an organization makes a wide range of changes in their organization *and* to the ERP software. As a result, in order to execute big-R reengineering, a firm will need substantial resources and time.

Example

Boeing is an example of a company that made both extensive organization changes and numerous software changes. Boeing surveyed the world of ERP software and found that none of the packages could meet their particular requirements. These requirements were so unique that BAAN software promised to work with Boeing in order to generate a version that was appropriate. However, Boeing did more than just change the software (Crow 1998); it also changed its own processes to reflect, for example, emerging concepts in "lean manufacturing."

As noted by an analyst at AMR, "[t]he size and scope of the project are truly unique" (Busse 1998). As of 1998, Boeing's new system linked 18,000 users across four geographical regions and nineteen different plants, with plans to increase that number to 50,000 users. Supporting the implementation required tremendous resources. For example, as noted by Busse (1998), Boeing had to maintain its 374 legacy applications in parallel with the new systems.

What was the payoff? As noted by one Boeing consultant, "[p]rior to ... [the new system] ... people had tunnel vision; now people see up and downstream" (Busse 1998). In addition, as noted by a Boeing vice-president, "[w]hen we ran the MRP [material requirements planning] engine the first time, we measured in terms of days. The same process today runs in 37 minutes."

Advantages and Disadvantages

There are at least two advantages of big-R reengineering. First, if successfully implemented, the firm gets the software and the processes it has identified as desirable, typically becoming a first mover. Second, in some cases (as discussed in the next section), the ERP "partner" (here, Boeing) incurs some of the costs and shares some of the implementation risk while becoming that first mover.

There can, however, be a number of disadvantages to big-R reengineering. Changing software can be a very expensive proposition. For example, as noted in Lamonica (1998),

Boeing got carried away with their changes, several thousand of them. It does become a maintenance nightmare – you end up building another organization to support the software. An upgrade of any sort is difficult, if not impossible.

Instead of changing the software, many firms are taking a position similar to that noted by Steve Cooper, director of enterprise systems at Corning:

Increasingly, what we want to do is to rely on PeopleSoft to do the bulk of the upgrade for us. If we've modified the software, it adversely impacts our ability to put the burden on PeopleSoft to do the upgrade.

As these remarks suggest, likely implementors of this big-R strategy are large firms with market power. Other candidates include cutting-edge firms that are seen by software vendors as offering an opportunity to gain a foothold in particular industries.

ERP Vendor Interest and Concerns in This Area

BAAN's sale of their ERP system to Boeing put BAAN on the map as one of the major players in the ERP business. Because of BAAN's addition of new best practices and software changes necessary for a large manufacturer and government contractor, similar firms (e.g., Litton Data Systems) were drawn to the BAAN software.

Alternatively, the more ERP software meets specific industry needs, the less it is likely to meet *other* industries' needs. As a result, making substantial changes in software or having specific best practices may actually constrain ERP sales to other firms. Some argue that BAAN's catering to Boeing gave it a system and an image that was not desirable to firms in unrelated industries.

Extensive Software Change:
Evolution from Big-R to Small-r Reengineering

In some cases of extensive software change, the ERP firm effectively partners with the implementing firm in an effort to expand the product capabilities and increase the set of available best practices. Accordingly, extensive software changes can result in industry-specific versions of the ERP software.

Special industry versions of software sometimes result from partnerships with the ERP vendors. In these settings, ERP developers are willing to shoulder some of the costs of software development so that other organizations can have access to the same processes. As a result, there is an evolution of ERP software: from software that could require big-R reengineering to that which would need only small-r reengineering.

Examples

Nash Finch

According to Stedman (1998), Nash Finch was one of the first to buy a version of SAP that was designed for retailers. When the implementation began, the necessary functionality for retailing did not exist but instead was being created as part of the project. As noted by the director of systems development at Jo-Ann Stores, "[w]here SAP retail and manufacturing functionality overlap, [SAP Retail] is very strong. But in areas that are specific to retail, it just doesn't have any depth."

Nash Finch spent $50 million before shelving the project to accommodate Y2K requirements. The problem was not with the financial modules but rather with other, retail-specific modules. As a result, Nash Finch indicated that they would continue the implementation after Y2K efforts were completed.

Nash's implementation is a big-R reengineering project. The current version of the software does not fully implement retail best practices. However, as Nash continues implementation of the retail version, those retail best practices will be captured and embedded in the software. Gradually, what starts as a big-R implementation for Nash Finch will generate software that can be used by other firms for small-r reengineering.

Boeing

As BAAN began to change their software to meet Boeing's needs, other firms in the defense industry began to express an interest in the same software. BAAN and Boeing worked together to advance the frontiers of encoded best practices. After those best practices became entrenched in the software, other users of those same best practices expressed interest in doing what for them might now be considered a small-r implementation of the software.

Implementation Failure and Success Factors

The highest probability of successful implementation of ERP software is when there is only a minimal need to change business processes and ERP software. This does not mean that all organizations should pursue choice of software with minimal change, but only that such implementation is more likely to be successful.

If most of the change is accounted for in changes to organizational processes (extensive change to organization processes and minimal change to software), then the lack of organizational adaptation, choice of the wrong best practices, or a resistance to change (among other factors) can lead to failure. Alternatively, if most of the change is accounted for by the software change (minimal

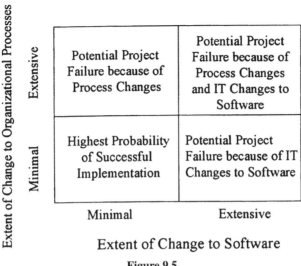

Extent of Change to Software

Figure 9.5

change to organization processes and extensive change to software), then the inability of the organization to implement large IT projects is among the factors that can lead to failure. Finally, if there is large change in the organization processes *and* in the software, then either or both sets of problems can affect the probability of success. See Figure 9.5.

Generally, for firms to reduce the potential of project failure requires resource expenditures and organizational adaptation in the area(s) that are likely to cause failure. Figure 9.5 provides guidance as to what kinds of resources and adaptation are necessary. For example, when there is minimal change to organizational processes but substantial software change, project management in information technology will become critical. In the converse scenario of extensive organizational process change and minimal software change, there will be a need for substantial "change management" in order to choose and implement the appropriate processes.

Which Quadrant? Which Approach?

Which approach – ranging from small-r to big-R – works best depends on the firm doing the implementation, their capabilities and constraints, and the trade-offs that they are willing to make. A firm may choose small-r reengineering and successfully implement the software yet miss the opportunity for reengineering their business processes. Alternatively, a firm may choose big-R reengineering, incurring large costs and implementation time, while becoming the first in the industry to install ERP and reap the attendant first-mover advantages. Throughout, a trade-off of costs and benefits occurs: one that is dependent on the criteria used (a) to make the ERP choice to begin with and (b) for its overall evaluation at the conclusion of the project.

Summary

This chapter has presented an analysis of the question, "Should business processes or ERP software be changed?" As part of this analysis we investigated the relevance of "as is" and "to be" modeling and found, for example, that in many cases an "as is" analysis is not necessary. In addition, this chapter introduced notions of "small r" and "big R" reengineering. We investigated those settings where major software change, major process change, neither, or both were likely to be appropriate. As with our previous discussions, there are no generally optimal solutions. Instead, the choice of approach is contingent on a number of organizational variables. Finally, we saw that using the specific industry information required for a big-R reengineering project by one firm can evolve into a situation where other firms will be able to do small-r reengineering using the same software.

References

Bailey, J. (1999). "Trash Haulers Are Taking Fancy Software to the Dump." *Wall Street Journal,* June 9.
Busse, T. (1998). "Boeing Takes Off with BAAN." *InfoWorld,* July 6.
Crow, B. (1998). "Integrating Lean Manufacturing Principles into BAAN ERP." Unpublished presentation, April 23.
Lamonica, M. (1998). "Customizing ERP Falls from Favor." *InfoWorld,* November 23, pp. 1, 57, 58.
Stedman, C. (1998). "Big Retail SAP Project Put on Ice." *Computerworld,* November 2, pp. 1, 104.

Appendix 9-1

Microsoft's SAP Implementation

Microsoft's customers and its internal product developers receive most of Microsoft's technology resources. As a result, the company's administration and

This appendix is based on "Microsoft Corporation," in B. Bashein, L. Markus, and J. Finley (1997), *Safety Nets: Secrets of Effective Information Technology Controls* (Morristown, NJ: Financial Executives Research Foundation). Many quotes in this case come from this reference. The material presented here has been specifically selected, ordered, rewritten, compressed, and focused.

internal management, including the finance group, often did not receive the information technology (IT) resources that they needed. According to Scott M. Boggs, assistant corporate controller: "One of the things that particularly struck me when I came here … was, rather than cutting edge, Microsoft financially was a Third World country" (Wallace 1995).

Microsoft

Microsoft Corporation was started as a partnership in 1975 and incorporated in 1981; it is headquartered in Redmond, Washington. Microsoft develops, manufactures, licenses, sells, and supports a wide range of software products, including operating systems (Windows), productivity systems (Word, Excel, etc.), development tools, reference tools (CD-ROM encyclopedia), and other products. Microsoft also markets books in support of their products, which are available for a wide range of computers, including PC-type and Apple computers. Microsoft is committed to a client server environment, where their products dominate the market. Microsoft is heavily involved in the research and development (R&D) for their products. The company has been very successful; it is debt-free and has significant cash on hand. In mid-1996 its market capitalization, larger than that of Intel and IBM combined, was estimated at over $78 billion.

Microsoft's PC software products are used worldwide. Most contractors and employees are computing experts who demand seven-days-a-week and "anyplace" access to Microsoft's computing resources. At the time of this case study, Microsoft had more than 40,000 clients, more than 3,000 servers, and over 20 terabytes of disk storage in the data center. In spite of their IT skills, Microsoft's internal IT users expected all their IT tools to be as easy to use as Microsoft's commercial software products. Because of their expertise and the software's ease of use, it probably is not surprising that Microsoft employees typically are intolerant of software training.

The Finance Group and Financial Systems at Microsoft

Microsoft's core business is technological innovation. As a result, decisions are based on technological and market considerations and not on sophisticated investment calculations. As noted by Boggs:

> The key risk for Microsoft is missing the windows for investment in a new technology. As a result, we are not a finance-driven company. We don't spend lots of time running cash flow analyses before an acquisition …. Our two business drivers are revenue and fixed cost – which is mainly people.

These requirements have influenced the finance group's role and capabilities. The finance group describes its two products as "relevant, timely, quality financial information" and "responsive, supportive, professional services." The IT roles and skills of Microsoft's finance specialists are difficult to distinguish from their IT support group. K. D. Hallman, director of the information technology group (ITG), noted that

> [t]here's a general passion for technology here. Some of our finance specialists, for example, got joint degrees in school [finance with MIS or computer science]. The finance and IT roles are constantly being redefined. Today, they're very blurry.

Microsoft's Legacy Systems

Microsoft, like many other large companies, had supported managerial and financial decision making by using a portfolio of legacy applications that had been built or bought during IT's mainframe and midrange eras. Microsoft saw these systems as an embarrassment. In addition, Microsoft's idea of using its own products internally was deeply embedded in company culture and operations.

In 1992, Microsoft's internal management information systems were extremely expensive to support. The year 2000 bug was a potential problem with many of the legacy systems. Microsoft had dozens of separate financial programs, each with specialized interfaces to common financial programs. Further, Microsoft's legacy systems were integrated with custom-designed interfaces that were difficult and expensive to maintain. Integrating legacy systems required labor-intensive custom programming. Changes in any one application or database would require additional extensive changes in other systems. Interface maintenance was roughly equivalent to application and database maintenance. Aidan Waine, NT SAP project director, commented as follows:

> We calculate that IT maintains over 400 applications. And these are big systems – to give you an idea of how big, we count MAC PAC [a major material requirements planning application suite] as one application. At least half of these applications can eventually be covered by SAP.

Some of the packages Microsoft had purchased had been modified so extensively that their vendors would no longer maintain them. Further, some important financial capabilities were not available, and financial systems were not meeting the needs for timely financial information. Moreover, current systems would not be able to support future growth projections. Microsoft's first close (of the financial "books") of the year took three weeks, and each subsequent month-end close took two weeks. Microsoft needed to speed data availability to management.

Problems with Processes and Artifacts

In addition to the legacy systems, there were some limitations and problems with basic financial processes at Microsoft – particularly purchasing. Scott Boggs explained: "We were not that efficient. For example, in purchasing we are sitting on huge piles of cash, so we never worked hard to negotiate purchasing discounts." Microsoft has a very loose purchasing policy, and almost all employees could make purchases up to $1,000 without prior approval (pre-approval was required for purchases over $1,000 and legal contracts). In addition to a loose policy, there was less than strict adherence to purchasing processes. As noted by Gregg Harmon, the SAP project director, "[i]n some cases we only found out about our [financial and legal] commitments when we received invoices."

There were problems with more than just the purchasing process.

- As noted by Harmon, Microsoft needed to improve its payment speed to vendors.
- In addition, Waine noted that "[i]n the fixed asset area, it … [takes] Microsoft 75 to 90 days to get an asset [e.g., computing hardware] on the books – after it was already on someone's desk."

There were also problems with system artifacts, such as vendor lists. For example, subsidiaries did not have a common chart of accounts or vendor lists.

The Decision to Replace the Internal Systems

In 1992, Microsoft began evaluating client server solutions for their internal management systems. The information technology group was responsible for the company's internal management and administrative systems. The ITG was centralized yet organized with applications development subunits for different internal customer groups.

The ITG had a policy of using package software whenever possible. They evaluated three different client server financial systems software: Dun and Bradstreet (D&B), Platinum, and SAP. D&B, still under development at the time of the decision, was chosen. Microsoft started a custom installation as a beta test site for the package. After about eighteen months, the implementation was stopped because the software was immature and there was a lack of business support.

In the meantime, SAP America had become the leading vendor in the United States for client server ERP software. SAP's R/3 did not yet have the full functionality of the mainframe-based R/2. However, SAP was the market leader, which meant that SAP would probably have the incentives and resources to provide future product enhancements.

In 1993, the ITG announced a plan for companywide implementation of the full SAP package. SAP modules for finance, operations, and human resources management were planned. This project went on for about two years until Bill Gates questioned the project's business goals and benefits.

After this project was terminated, the ITG was reorganized. After the reorganization, applications developers were spun off to their internal customers, so that the finance group had its own finance IT (FIT) group. In addition, John Connors (at the time, corporate controller) was named Microsoft's chief information officer. Connors's charge was to improve processes and reduce costs. In addition, Connors was to ensure that Microsoft's executives understood the value of the company's internal investments in IT.

SAP Choice

Executive commitment to installation of SAP's R/3 financial modules occurred in 1994. Steve Ballmer (executive vice-president of sales and support) had been visiting Wal-Mart, a company for which Ballmer had taken personal responsibility. Wal-Mart was widely known for its sophisticated information systems, and Ballmer was impressed with what they had accomplished.

Connors described Ballmer's reaction to Wal-Mart systems, and his charge for Microsoft's finance group:

> We had just had a very bad budget process. ITG and finance had developed a new budget tool and it didn't work .... It was not fun .... I was just back from vacation, and Steve Ballmer was just back from Wal-Mart. Steve knocked and opened my door. I knew it was Steve; he has a really distinctive knock. He walked in and said, "You guys [expletive deleted]." I got the message.

Assistant corporate controller Scott M. Boggs, the business lead for the project, noted: "[w]e didn't do a real financial justification of this project. It's strategically important and has the support of top management."

SAP and the Project Scope

The scope of the project included the set of financial processes from initiation of financial transactions (e.g. purchases) to the publication of financial results, for internal use by managers and finance staff and externally by investors. To improve and enable this process, finance and FIT decided on SAP's R/3 financial modules (excluding treasury, tax, and audit applications). SAP AG is one of Microsoft's strategic business partners, and SAP's R/3 was the leading client server–based enterprise software package in the United States. It had the ability to handle multiple currencies – a major plus, since Microsoft has subsidiaries in 60 countries – and there were "best practices" built into the package. In addition, SAP and Microsoft jointly market a technology package consisting of the R/3 software, Microsoft's Windows NT network operating system, and Microsoft's SQL Server database management system.

As of 1995, however, most large companies that selected SAP R/3 used the UNIX operating system and Oracle's database server. At that time there were no examples of SAP and Microsoft technologies for more than 1,000 users with multiple sites. Gregg Harmon estimated that "[t]he largest single site NT installation [before ours] was 650 people; and the largest multiple site was 350. And they only work days." Using a beta version of Windows NT 4.0 (which was released to the public in August 1996), the team's plan was to build an SAP/NT/SQL Server installation for an estimated 2,000 (nonconcurrent) users.

Project Team and Schedule

At Ballmer's suggestion, John Connors championed the financial systems replacement project. He described his role as "the strongest advocate of the . . . project. I sold it to BOOP [Bill and the office of the president]." Connors described the key role he played in the new project:

> I'm not a project manager. I'm not a developer. My job is to remove obstacles to progress. That means several things. First, it means getting ITG to understand that they had to be part of a team. I made sure that the business lead for the project was a finance person. That was a change from the way they [ITG] had done things in the past. Second, it means saying "this is what we're doing . . . . You don't have a say in whether we're doing this, just in how." I took a "short long" trip around the world to communicate this to everyone. Third, it means gaining executive support – getting Ballmer or Bill to nuke any obstacles that arise. Fourth, it means telling everyone how important this is to Microsoft and teaming with marketing and sales. Fifth, it means hiring a very effective project manager.

Connors put Scott Boggs in charge of the overall project. Boggs was not an IT person, but he was an important figure in the business process directly affected by the project, the "process owner." Microsoft also called it "business lead" – someone who benefits if the project succeeds and suffers if the project fails.

Boggs and Steve Lindemann (senior consolidations manager) attended an SAP users' conference to learn more about implementation challenges. As noted by Lindemann,

> I feel fortunate that I was involved in the project from the beginning. It would have been easy for accounting to leave it all to information services. Scott and I went to an SAP annual conference last year, and every company said the same thing. [Don't leave it to IT.] We looked at each other and decided that we were not going to make that mistake.

The users' conference made it clear that R/3 was very complex. It was also apparent that it was useful to have people with prior R/3 experience implementing the package. As a result, Connors hired Gregg J. Harmon from a major IT consulting firm as SAP project director. Aidan Waine from ITG was chosen as NT SAP project manager. These dual project managers shared an office and divided responsibilities using consensus.

The project leaders were given an extremely aggressive schedule. A goal was set to complete the project in nine months. There was a lean budget, but managers were given license to use whatever internal human resources they needed to do the project. Project managers directly interviewed and selected project personnel. However, the thinly staffed finance organization could not afford to replace the people selected for the project. As a result, project team members were expected to do two full-time jobs. This worked most of the time, although, as noted by one project team member, "during the [month-end] close, we'd often lose them."

Microsoft project team members also worked with several groups of consultants and vendors. As Harmon remarked,

> [v]endors are necessary, but it's not good to rely on them too heavily .... There are trade-offs for companies using consultants to do SAP installations. You get a fast start with consultants [because they are already trained on the package]. But the risk is that your own people aren't trained to do the maintenance.

Microsoft project team members worked with consultants and hired outsiders who already had SAP expertise in order to move quickly and train their own people on the package. Microsoft's SAP implementation effort eventually involved roughly equal numbers of finance specialists, IT people, and external consultants.

Small-r Reengineering

Microsoft decided to implement the SAP package with no modifications. Microsoft wanted to capture the benefits of the best practices that had been built into the package. Aidan Waine referred to this approach as "small-r reengineering with SAP R/3 as change agent." According to Waine,

> [t]his company is in a constant state of flux. You might even say we have chronic ADD [attention deficit disorder]. We have no luxury to engineer our processes carefully. Our basic project philosophy is to get things out fast and get better as we go. Everything is timed for 12-month delivery.

Because Microsoft did not customize SAP, they could estimate more accurately the time required to install the package. However, despite their decision not to modify SAP, the project was extensive: 3,000 tables of parameters capturing Microsoft's business arrangements had to be established. Although the SAP project dramatically reduced the number of discrete financial applications (from 24 to 5), software interfaces were needed to link non-SAP financial programs to the SAP database.

Sticking to the decision not to customize SAP was hard. The organization model used in SAP did not correspond to the model used at Microsoft. For instance, the term "department" is not used in SAP. Instead, R/3 is written in terms of "profit centers" and "cost centers" (each identified by a code). Because the project team did not want to modify the package, all of Microsoft's departments had to be redefined as profit centers or cost centers when the SAP package was installed.

In addition, SAP uses different business models than those employed by Microsoft. For example, SAP employs the concept of "internal orders" for intraorganizational transactions, which are common in Microsoft's matrixed business model. Microsoft's required internal orders pertained to departments, products, and projects. Finance wanted to use three fields to classify its transactions, but internal orders in SAP were represented using only a single field. Accommodating Microsoft's existing business model would have required modifying the SAP package. But the alternative involved creating unique codes for each one-, two-, or three-way combination of the business entities, a combinatorial nightmare that would be virtually impossible for business unit financial staff to remember. In addition, internal order codes would be constantly changing. Further, if code tables were maintained at headquarters, field personnel could also face substantial delays in their ability to enter transactions. As a result, in order to make the scheme of single-field, internal order number work, the project team had to allow subsidiaries the ability to create new combinations as required.

Unfortunately, such implementation compromises reduced the standardization that Microsoft was trying to introduce into its worldwide chart of accounts. Steve Lindemann reported that

[i]t was a very painful decision. It used to keep me up at night. It made me sick. The way we did it was the least terrible option. It does meet the business needs, but it's not the prettiest solution. I was concerned about the people who had to work with all these codes. The next version of R/3 gives more flexibility in the handling of internal orders. Will we change over to the new version? We haven't crossed that bridge yet either. It's a very complicated decision. Change would be difficult.

Training

In spite of their efforts to integrate their own personnel into the development process, Microsoft had to make considerable increases in training budgets. SAP's R/3 required a substantial amount of training. As a result, the SAP project team redesigned SAP's training to be as short and fast-paced as possible. The new training integrated business knowledge with SAP skills building. According to Lindemann, "I've taught a six-hour course on entering journal entries into SAP 20 times now. It was as much a general ledger course as a systems tool course."

SAP training was required for those employees that would use SAPSET (standard expense tracking). Microsoft threatened to turn off the accounts of those who did not go to training. This approach apparently worked; as noted by Steve Lindemann, "[w]e only had to turn off 20 or so accounts."

Implementation

In August 1996 – nine months after the project started – the SAP installation was 90% complete, with about 1,000 hands-on users at corporate headquarters. Gregg Harmon remarked that "[w]e hit all our milestones. This is the biggest ITG project done well in Microsoft to date."

The first-ever SAP "first close" looked like it would take the same three weeks required of the legacy systems. However, Microsoft expected that future first closes of the year could be shortened substantially; in addition, they expected that subsequent monthly closes could be shortened to two or three days.

Out of Microsoft's 20,000+ employees, there were expected to be about 2,000 worldwide accounting and finance users, mostly for business units or products. Of course, many others were expected to use the data produced by R/3.

Open Issues

Although the implementation was considered a success, there were a number of open issues that still needed to be addressed by the project team.

International Rollout

Beginning in August 1996 and continuing over the next twelve months, Microsoft planned to roll out SAP to international subsidiaries. Most of the firm would be using virtually the same model. As noted by Waine,

> No one has said that SAP is worse than the old system. The subs have been pretty much starved for capability. And they've had input into the SAP installation for a year. Some subsidiaries think they can do it better by themselves; they're quite autonomous. But we're not giving them the resources to make major changes. Microsoft is pretty autocratic when it comes to reporting to the center, especially in finance. We've given them 33 tables that they can configure [to reflect their local needs and different ways of doing business]. Here [at headquarters] we're configuring 3,000 tables.

Human Resources

Human resources is the next area of legacy applications slated for replacement. Although CIO John Connors describes himself as "a big supporter of SAP," Connors believes that SAP's human resources applications are not as strong as they should be for Microsoft's needs. Further, human resources personnel prefer PeopleSoft to SAP. Connors noted, "I'm going to SAPPHIRE [SAP's user conference] next week. I want to spend some cycles with them learning about where they're going with HR."

No Changes to Purchasing Processes

Some finance group members expressed the concern that there was no change to the internal purchasing process authorizations. For example, there was no change in the policy that anyone can spend up to $1,000 without authorization. The primary control was managers monitoring their budgets. Gregg Harmon explained that the risk was relatively small: "Seventy percent of our purchase transactions are under $1,000, but they only account for 2% of our spending. If anything, we've loosened controls on spending ...."

What Made It Work?

Scott Boggs is often asked for his advice on IT implementation of SAP. He explains his response as follows:

> I talk to lots of other companies about what we're doing here. They always ask my advice. Lots of times I don't know what to tell them. They say things like: "We don't have a network. Half of our people use this, half use that. The CEO doesn't use IT and doesn't think it's important. What should we do?" I tell them, "I don't know. I don't have those problems."

Reference

Wallace, P. (1995). "Microsoft's Finance Department Gets Up to Speed." *InfoWorld,*
June 5, p. 58.

Questions

1. What risks are there in this engagement for Microsoft?
2. Why did the first two ERP implementations not work?
3. How important is it to get the business areas involved in ERP systems?
4. Explain "small r" reengineering. What is likely to be "big R" reengineering?
5. Consider the implementation compromises. Were too many made? Why or why not?
6. Was the engagement a success? Why or why not?
7. Should Microsoft implement SAP's HR package? Why or why not?

10

Designing ERP Systems

Choosing Standard Models, Artifacts, and Processes

One of the critical sets of decisions that needs to be made in any ERP concerns the choice of common standard MAPs – that is, models (e.g., organization models), artifacts (e.g., vendor numbering schemes and lists), and processes (e.g., order management). Two expressions, "common ... and global" and "input by many," have received attention as guidelines to help implement those common MAPs in some companies (see CIO 1996). Rather than following the historical process of each business unit choosing its own MAPs, as is often done in legacy systems, an ERP-structured company will use the same MAPs for each business unit. However, the choice, implementation, and use of common MAPs is not easy. Within a given firm there are likely to be conflicts over those MAPs. As a result, this chapter addresses the following questions.

- Why are MAPs important?
- Where do common MAPs come from?
- Why do firms need common MAPs for ERP systems?
- Why didn't firms have common MAPs prior to ERP systems?
- Why is it difficult to choose common standards?
- What are some choice motivations?
- In order to develop "common ... and global" processes, what methods have been used to choose common ERP models, artifacts, and processes so that there is "input by many"?

Why Are MAPs Important?

The quality of the MAPs will have a huge impact on the overall success of the ERP implementation. MAPs that are not efficient or effective for a particular firm can drag down its overall performance. Similarly, MAPs that meet the needs of a firm can push it to better performance, giving it a competitive edge. As a result, the process that firms use to choose MAPs is important.

Where Do MAPs Come From?

Where do the standard artifacts and processes come from? This section summarizes interviews from two large firms concerning the source of their common MAPs.

An interview with an ERP project manager at Nestlé (United States) found how Nestlé generated different business process options for its SAP implementation. First, Nestlé decided that it would implement common MAPs in all three of its U.S. divisions. Second, each of the three divisions' existing MAPs became candidates that would be evaluated. Third, both SAP and the advising consultant's best practices databases were used to generate candidate MAPs. In some cases, hybrid MAPs were developed, based on multiple sources of information. Fourth, a multifunctional team used both sets of inputs to decide on company standard artifacts and business processes.

An interview with the deputy ERP project manager at Litton Data Systems found that a similar approach was used to generate the portfolio of standards for its BAAN implementation, but that the source of standards was somewhat different because of specific industry considerations. As with Nestlé, a multiple function and multiple departmental group was founded to generate and evaluate the choices. However, since Litton is involved in a number of projects for the federal government (in addition to other sources), federal-based MAPs were also necessary in order to meet mandated guidelines (e.g., reporting requirements for government contracts).

Why Didn't Firms Have Common MAPs Prior to ERP Systems?

Why does ERP require common MAPs now, and why didn't firms have common MAPs prior to ERP systems? The answer derives from a number of factors that include technology, exploitation of local differences, and divisional control. Historically, much corporate software has been customized and local. Technology limitations meant that divisions could make different choices in hardware and thus in (hardware-dependent) software. As a result, individual divisions and plants exploited unique characteristics by choosing software and MAPs that met their divisional needs. For example, artifacts (e.g., vendor lists, product lists, customer lists, charts of accounts) were chosen and developed with *local* processes and decision-making needs in mind.

Disparate software and MAPs made it difficult to integrate across multiple divisions. It therefore became increasingly difficult to coordinate divisions to meet increasing global customer needs and competition. However, client server technology, networking technology, and software were gradually becoming available that would allow the choice of common ERP software as an enterprise solution.

Why Do Firms Need Common MAPs for ERP Systems?

Unlike customized local software, ERP software is "packaged" – designed and developed to facilitate implementation in many companies and, moreover, to allow integration across all divisions and plants in a company, facilitating a common corporate view. As noted by Brownlee (1996, p. R18), "When Colgate employees log on to the network, the same menu options appear on their screens, regardless of whether they're in Cambridge or Burlington or New York."

Thus, ERP software enables common MAPs. But why do firms need to adopt common MAPs when they implement an ERP system? This section summarizes a number of reasons gathered from a range of ERP implementations:

- the software requires it;
- improved customer response;
- regaining control of processes;
- need for a common view of the organization;
- value creation and cost reduction.

The Software Requires It

Perhaps the most immediate reason that standard ERP artifacts and processes are adopted is because ERP software requires it. As noted by a vice-president of Red Pepper Software, "[w]hether it is steel, rubber, or electronics, with SAP it is a common set of parameters that drive individual plants" (Vaughn 1996, p. 72). The need for standard product lists and standard price lists also was seen in the SAP implementation by Owens Corning, which

> traditionally had operated as a collection of autonomous fiefs. "Each plant had its own product lines," says Domenico Cecere, president of the roofing and asphalt units. Each plant also had its own pricing schedules, built up over the years of cutting unique deals with customers .... [SAP's] R/3, however, effectively *demanded* that Mr. Cecere's staff come up with a single product list and a single price list. (White, Clark, and Ascarelli 1997; emphasis added)

Improved Customer Response

Enterprise resource planning software allows integration of multiple divisions and plants within an enterprise, ensuring that all users have access to the same information. This improves the company's responsiveness to customer needs. Owens Corning chose to implement SAP's R/3 to facilitate a single view of the company and to allow quick response to customers (White et al. 1997).

> Up until now, customers called an Owens Corning shingle plant to get a load of shingles, placed a separate call to order siding, and another call to order the company's well-known pink insulation.
>
> [The company's new vision was that] Owens Corning should offer one-call shopping for all the exterior siding, insulation, pipes, and roofing material that builders need. [SAP's] R/3 will give Owens Corning the ability to make that happen by allowing sales people to see what is available at any plant or warehouse and quickly assemble orders for customers.

Regain Control of Processes

In some cases, different organizational locations evolve processes that appear to get out of control. In order to get those processes back under control and to facilitate organizational adoption of best practices, firms adopt ERP systems. As seen at Vandelay (McAfee and Upton 1996, pp. 4–5):

> Vandelay's sites' operations practices were as varied as their information systems. There was no uniformly recognized "best" way to invoice customers, close the accounts at month end, reserve warehouse inventory for a customer order, or carry out any of the hundreds of other activities in the production process that required computer usage or input.
>
> To alleviate ... problems with systems and practices, Vandelay decided to purchase and install a single ERP system, which would incorporate the functions of all the previously fragmented software. The company would also standardize practices across sites.

Common View of the Data

Some firms attempt to develop ERP implementations with each business unit having their own set of artifacts. However, most firms that choose ERP want to avoid the negative impacts of diverse and fragmented MAPs.

In order to generate a common view of the data, a common set of organizing artifacts (e.g., vendor lists and customer lists) is required. Those artifacts provide the ability to gather information in disparate settings yet still retain a common view across each of those settings. For example, as noted by Vaughn (1996, p. 74):

> Elf Atochem North America Inc., Philadelphia ... is moving 13 business units over to SAP software .... Elf Atochem came to SAP because its various companies had been reorganized to work as one. [As a result, the company] had inherited "a lot of different computer systems, a lot of different ways of doing business, and a lot of hand-offs." A common view of diverse data was important ....

Value Creation and Cost Reduction

Implementing standard artifacts and processes can create value and reduce costs. For example, as noted by Pirelli's director of information technology Arrigo

Andreoni, "[t]he more standardization there is, the easier it is to implement new ideas and respond to new opportunities" (Wakin 1998, p. 48). Andreoni also notes that standardization can reduce costs. As an example, before standardization Pirelli had a full-service back office and customized software in each of five countries. Enterprise resource planning software was used to replace the multiple back-office staffs with a single back-office staff in Switzerland, cutting costs by 25%. Now offices in each country send data to Pirelli's central servers.

Why Is It Difficult to Choose Common Standards?

Because ERP systems are developed for the enterprise, an ERP implementation requires many enterprise decisions, including which standard artifacts and processes will be used. For example, in the case of Vandelay (McAfee and Upton 1996), a number of standards were adopted – including a common chart of accounts, common vendor numbers, and common part numbers – in order to facilitate communication and coordination between different business units such as factories and divisions. However, some MAPs benefit particular business units differentially. What is adopted for global usage is not always what is best or preferred locally at each of the divisional locations. For example, Vaughn (1996, p. 72) quotes Chris Roon, a vice-president of Red Pepper Software, who admits that standard ERP artifacts are "useful where financial viewers want to consolidate information across diverse operating units, but … the common view may not be optimum for individual divisions." When ERP standardizes processes and artifacts there are global benefits, but some of them come at the cost of sacrificing local customized capabilities.

Moreover, because of differential corporate and divisional benefits that arise from standard artifact and process choices, group decision making can become a political process that is subject to strategic behavior designed to advance each faction's cause. Thus, differences in interests between global and division views can ultimately result in conflicts. Also, choices of standard artifacts and processes do not always maximize a corporation's overall utility, especially when some divisions enjoy more benefits than others.

Product Lists: A Common Standard Artifact

As an example of a common standard artifact, consider the product list. A product list gives a unique identifier to each product for purposes of entering and reporting data about that product. If a common product list is adopted across all divisions, this means that each division uses the same product list, which typically provides sufficient flexibility to add new products in addition to encompassing all the existing products. However, information needs are not necessarily the same in each division of a company, since divisions have different products and markets. Some products are likely to be used in only one division; some divisions are likely to need more extensive and detailed product lists than others; and so on.

As a result, the various business divisions and units each experience a different set of costs when implementing a standard product list, depending on whose product list was chosen. If a particular division's product list were used as the common standard across the company then that division would naturally face minimal training and implementation constraints; further, such a list would fit its interests better in terms, for example, of facilitating the addition of new products. However, for other divisions it would be an entirely new product list that requires training and may not fit that division's needs.

Divisions usually find that the common product list is not as efficient as their previous one. For example, items from a product list coded with twice as many digits can take twice as long to enter and have a higher probability of data entry errors, suggesting that the quality of the information could decrease.

How Much Variance in Standards Is There between Corporate and Divisions?

What percentage of the MAPs are made available for divisional configuration? The share probably varies from organization to organization. However, as noted in a discussion about Owens Corning (CIO 1996), there are always exceptions to "common … and global" principles: "You'll always have variations in processes at the business unit level with SAP, particularly around customers … it's my job to make sure the variations are the exception rather than the rule." In Microsoft's implementation, headquarters configured 3,000 tables whereas local offices were given roughly 30 tables (Bashein, Markus, and Finley 1997).

How Seriously Do Divisions and Corporate Take the Choice Process?

The choice of common MAPs is serious business for all involved. In some cases, the choice processes can become very emotional. For example, as seen in the SAP implementation by Owens Corning (White et al. 1997), there was substantial conflict over an artifact:

> R/3, however, effectively demanded that Mr. Cecere's staff come up with a single product list and a single price list. The staff initially fought ceding control over pricing and marketing to a computer-wielding central command. "My team would have killed if we'd let them," he says.

As a result, it can be important for firms to try to choose MAPs using a systematic approach.

What Are Some Choice Motivations?

Enterprise resource planning committees generally include representatives from each of the affected divisions, which can express their preferences on adopted standards through their representatives on these committees. Divisional preferences may be driven by any of a wide range of motives, as follows.

Maximizing Corporate Benefit. Ideally, each division will put aside its individual goals and work toward developing a set of standards that maximizes the global corporate benefit. Realistically, divisions have been known to make self-serving decisions.

Minimizing Divisional Change Costs. Rather than maximizing corporate benefit, a division may work to keep its own artifacts or processes in order to minimize change costs – such as training current (or hiring new) personnel to use (or implement) the new standards. A division that is unsuccessful in promoting its own standards as standards for the company can still minimize its change costs by choosing the standard that is closest to their own.

Responding to Competition. Global competition has forced many companies to outsource some activities formerly done by corporate divisions. A division thus faced with potential extinction may respond with extensive reengineering. Accordingly, in such settings we might see divisions work to implement substantial change in standard artifacts and processes (see e.g. Hammer 1990).

These varying motives suggest that there are incentives for divisions to induce the corporation to choose standards that benefit particular divisions. As a result, companies need to think about how to choose these common and global standards.

Methods for Choosing between Standards

How do organizations implementing an ERP system make the choice between common artifacts and processes? As just noted, there are a number of sources of ERP standards, and the cost to individual divisions will vary depending on which standards are adopted.

How do organizations choose which set of standards (e.g., vendor lists or product lists) to adopt? As noted by McAfee and Upton (1996), companies have some questions as to how to choose standards for ERP systems. For example, a project manager is quoted as having a strong bias toward "input by many, design by few" but did not know how to put that statement into practice. How do we gather input by many? What group choice rules have firms used?

Majority Vote

According to Wakin (1998), Pirelli's director of information technology uses what he calls "democratic governance" across divisions to achieve the standardization in core business areas required for ERP implementation. Although democratic governance can take many forms, it is suggestive of a "majority vote" approach – whichever standard the majority of divisions prefers becomes the standard. A majority vote choice rule is one where the company adopts MAP x instead of MAP y if more divisions prefer MAP x to MAP y.

Borda Rule

An alternative way to gather input from many is to (a) have panel members vote by ranking the options and then (b) sum the number of first-place (second-place, etc.) rankings received by each option. For example, if there are five different ways to execute the business process "reserve inventory for clients," then each panel member would rank each of the alternatives from 1 to 5. Each vote for a ranking of 1 gets five points (for a ranking of 2, four points,...); the option that totals the most points is, by the *Borda rule* (see e.g. Fishburn 1971), the winner.

An interview with the deputy project director of the Litton Data Systems PeopleSoft implementation found that they used a Borda-like approach. First, Litton defined "key aerospace and defense features and thrusts," including:

(1) defense business processes –
  • government forms,
  • progress payments,
  • moving average actual costs;
(2) financial management –
  • division budgeting and resources planning,
  • cost schedule control,
  • multiple rates for labor and burden.

Second, given these and other requirements, different ERP vendors were evaluated and ranked on their ability to provide each feature, with points given for the rankings earned (as just described). Third, the choice of ERP system was made based on the final ranking as determined by total points.

Pareto Optimality

As noted previously, Pirelli's director of information technology uses what he calls "democratic governance" to achieve standardization in core business areas required for ERP implementation (Wakin 1998). Pareto optimality is a criterion for choosing between pairs of ERP standards by first assessing the preferences of each division. Of course, if all business units agree on process x over y then process x is chosen. In the more usual case where divisions do not agree on the preferability of different processes, the "Pareto optimal" solution is the choice whereby no *other* choice could (a) satisfy any one division more without (b) satisfying any other division less.

Dictatorship

In other settings it is apparent that, rather than a voting approach, the solution used in choosing standards for an ERP system is closer to a dictatorship. For example, a discussion of Owens Corning notes that "it would be relatively easy for the Toledo-based team to huddle and make SAP process choices for the company unilaterally" (CIO 1996).

Another example (Brownlee 1996) is Colgate, who implemented a network of "supplier managed" inventories (for both its customers and suppliers) using SAP. Colgate provided a number of its most important suppliers with SAP so that they could directly access Colgate's SAP system. As part of the trade-off, Colgate does not have to pay for product ingredients until it actually uses them. In this setting, Colgate specified many of the business processes, provided standards in terms of various pricing lists, and so forth.

A division is likely to be a "dictator" if it is the corporate office division with a particular view of the world, or a large division with substantial relative power.

Which Is the Best Method?

The choice of MAPs can have a positive impact on a firm's performance; hence, their choice is critical. Although there are a number of rational approaches, no one approach is best in every circumstance. Further, it can be shown that each approach fails in some situation. As a result, the choice of those MAPs is often a political process.

Summary

Prior to ERP systems, firms generally did not employ common and global models, artifacts, and processes (MAPs). However, implementing ERP systems generally requires a set of common MAPs for a number of reasons: system requirements, customer service, process control, a corporate-wide database, value creation, and cost reduction.

Companies typically generate a portfolio of standards by choosing among existing company MAPs (as found in different divisions), consultant and ERP best practices, and hybrid approaches. Unfortunately, it is not usually clear which of a portfolio of artifacts and standards should be adopted, particularly since any choice is likely to benefit certain divisions more than others. The standardization of MAPs can be quite extensive, and divisions may be allowed to choose as few as about 1% of the MAPs.

Owing to the limited number of MAP choices and their importance to individual divisions, the choice of MAPs can result in emotional conflicts. It is therefore important to examine rational approaches that have been successfully used by other firms. Examples include majority rule, the Borda rule, Pareto optimality, and dictatorship.

References

Bashein, B., Markus, L., and Finley, J. (1997). *Safety Nets: Secrets of Effective Information Technology Controls.* Morristown, NJ: Financial Executives Research Foundation.

Brownlee, L. (1996). "Overhaul." *Wall Street Journal,* November 18.

CIO (1996). "SAP Key Roles." ⟨www.cio.com/forums/061596_sap_roles.html⟩.

Fishburn, P. (1971). "A Comparative Analysis of Group Decision Methods." *Behavior Science,* November, pp. 538–44.

Hammer, M. (1990). "Reengineering Work: Don't Automate, Obliterate." *Harvard Business Review,* July/August, pp. 104–12.

McAfee, A., and Upton, D. (1996). "Vandelay Industries." Report no. 9-697-037, Harvard Business School, Cambridge, MA.

Vaughn, J. (1996). "Enterprise Applications." *Software Magazine,* May.

Wakin, E. (1998). "Global Strategies Drive Pirelli." *Beyond Computing,* January/February, pp. 46–8.

White, J., Clark, D., and Ascarelli, S. (1997). "This German Software is Complex, Expensive and Wildly Popular." *Wall Street Journal,* March 17, p. 1.

11

Implementing ERP Systems

Big Bang versus Phased

"Phased" and "big bang" are the two primary (and contrasting) approaches used to implement ERP systems. This chapter investigates what these terms mean, some properties of each approach, and some of the advantages and disadvantages of each. This chapter also analyzes the choice of the implementation methodology in light of organization size, complexity, and structure and in terms of the overall extent of the implementation. Finally, some additional terms and approaches used to implement ERP systems are discussed briefly.

What Is a Big-Bang Implementation?

In a full big-bang implementation, an entire suite of ERP applications is implemented at all locations at the same time. Using big bang, the system goes from being a test version to being the actual system used to capture transactions in only a matter of days (hence the name "big bang"). Big bang requires simultaneous implementation of multiple modules. In the case of Quantum (see the appendices of this chapter), this meant implementing 17 modules of Oracle applications at 23 sites around the world over a period of about one week. Under this scenario, big bang requires a large amount of testing before cutting over from legacy systems to the new system. In the case of Quantum, this meant six to eight months of system testing prior to going "live" and actually using it to process transactions.

The big-bang approach usually employs a three-step process. In the first step, virtually all relevant processes and artifacts are chosen (or developed) and implemented in the software. For Quantum, this stage extended from roughly January 1995 through August 1995. In the second step, all modules are tested individually and for their interfaces with other modules. At Quantum, this second stage took roughly from September 1995 through April 1996. Problems found during the testing stage constitute feedback that is used to further develop and fix the modules. In the third step, the old system is turned off and the new system is turned on. After the implementation there are always minor changes that need attention; however, because of the extensive testing, it is hoped that no major changes are needed.

What Is a Phased Implementation?

A phased approach is one where modules are implemented one at a time or in a group of modules, often a single location at a time. Phased implementations are sequential implementations that consist of designing, developing, testing, and installing different modules. Unlike big bang, phased implementations require that substantial attention and maintenance be given to legacy systems in order – at each phase – to facilitate integration with the new ERP system.

As compared with big bang, the phased approach has smaller "slices" of module process and artifact design, development, testing, and implementation. Siemens, for example, used a three-phase SAP implementation (Hirt and Swanson 1998). In the first phase, Siemens implemented the finance, controlling, accounts receivable, accounts payable, and purchasing modules. Phase-one design, development, testing, and implementation was to be completed over the period October 1995 to September 1996. Phase two was to include much of the materials management, production planning, and quality planning modules, taking place over the period October 1996 to April 1997. Finally, the remainder of the modules were set to be designed, developed, tested, and implemented over the period April 1997 to September 1997.

What Are the Advantages of a Big Bang Approach?

No Need for Temporary Interfaces

Big-bang implementations basically leave existing legacy systems intact until they are replaced. Because it replaces a number of legacy systems all at once, big bang does not require temporary interfaces. For implementations with numerous ties to legacy systems, substantial resources may be required if a big-bang approach is *not* used.

Limited Need to Maintain and Revise Legacy Software

Since big bang entails an immediate migration from the legacy systems to the new ERP system, there is little need to spend time or resources maintaining or changing the legacy systems. As a result, with the big-bang approach, virtually all maintenance and development resources required under a phased approach can be devoted to development and testing of the new system.

Lower Risks

In a phased implementation project, team members participate in phases as different modules are implemented. In a big-bang implementation, however, the entire project team attacks the project at roughly the same time. As a result, some (such as Hank Delevati, CIO at Quantum) have argued that a big-bang approach has lower risk than a phased approach: "The phased approach is ... riskier, because you won't get everyone involved and coordinated" (Radosevich 1997). In addition, the risk of losing employees before completing the engagement is lower with big bang. Experience of engagement participants is not "whole" until they complete the implementation. As a result, with big bang, participants are more likely to stay with the firm for the entire effort.

Functionality Linkage

Desired functionality may require that multiple modules be linked. In this case, such modules must be implemented before the functional features are available ("available to promise" is one such feature). Since big bang implements all modules at once throughout an organization, linking modules can be completed more rapidly and so users have less of a wait for cross-module functionality.

No Going Back

With big bang there is no legacy system to go back to. As a result, firms must forge ahead with the new system even if it is not comfortable to do so. Knowing that there is no return makes it easier to not look back. It can also facilitate a commitment to the system that might not otherwise be possible.

Shorter Implementation Time

One of the reasons that ERP engagements fail is that they take too long (often 1–3 years) to complete. Lengthy implementation can cause engagement failure for a number of reasons. The longer an implementation takes: (a) the more the requirements will change; (b) the more likely that personnel involved will turn over; and (c) the longer that project opponents will have to work against the project.

Although implementation time can almost always be influenced by the amount of resources devoted to accomplishing the task, neither phased nor big-bang necessarily has a longer total "person year" implementation time. However, because the big-bang approach handles design, development, testing, and implementation of all modules simultaneously, it generally has a shorter duration from start to finish. Big-bang implementations may take less time also because no temporary interfaces to legacy systems need be built (as with phased systems).

Cost

Which approach is more costly? If all goes well and there are no surprises, then the big-bang approach would probably have the lowest out-of-pocket expenses because there is (at most) limited work on existing legacy systems and temporary interfaces.

What Are the Advantages of Phased Implementations?

Peak Resource Requirements Are Less Than with Big Bang

Some firms, such as Siemens (Hirt and Swanson 1998), have chosen a phased approach because of limited resources. Whereas a big-bang approach requires substantial one-time resource use, a phased approach can spread those peak resource requirements over multiple phases. With the phased approach, the resources required for any given phase can be limited and made feasible for even the most constrained organization.

More Resources Can be Devoted to a Particular Module

If necessary, virtually all of a firm's design, development, and testing resources can be focused on one particular phase and its associated modules. This is in contrast to a big-bang approach, where resources must be spread over multiple modules simultaneously. This issue could be critical to firms with tight resource constraints or limited organizational slack. Without the ability to muster the appropriate level of resources, the project can face substantial risk of failure.

Lower Risks

Big bang is an "all or nothing" approach. One malfunctioning module in a big-bang implementation can cause the implementation to fail. As a result, some firms may feel that the potential for total system failure is just too high with a big-bang approach, or they may be uncomfortable with the all-or-nothing nature of the "bet" being placed. A phased strategy is generally regarded as a lower-level risk.

Legacy System Fallback

In a big-bang implementation, the legacy system is turned off. This means there is no alternative if (because, say, one piece fails) the whole system collapses. Because a phased approach installs the ERP only one piece at a time, it allows an organization greater ability to ensure that a module works before the alternative is turned off. A phased approach is thus more conservative since it provides a back-up.

Personnel Gain Knowledge in Each Phase

In a phased implementation, knowledge gained in one phase can be transferred to other phases; personnel in one phase can use new knowledge gained from the implementation in later phases. As a result, modules can be designed by increasingly more experienced people as design, development, testing, and implementation issues are fed back to the project team. This advantage is less of an issue if much of the work is being done by external consultants, although they too may gain knowledge from one phase for use in another (even if that knowledge is specific to a particular organization).

Project Managers Can Demonstrate a Working System

In a phased approach, successful implementation of a module can be used to show the rest of the company that the system works. Generating successful results may be necessary in settings where initially there is limited organizational or top-management support. Success in one phase can be used to demonstrate to management that the system can and will work.

If this is a primary reason for a firm to use a phased approach, we are likely to find that the "easiest" modules are likely to be implemented first. For example, those modules requiring only minimal reengineering would be the first implemented. This can be seen in the case of Siemens (Hirt and Swanson 1998), where the first package implemented had almost complete conformance between its capabilities and the firm's requirements.

Time between Development and Use Is Reduced

Unfortunately, with a big-bang approach there can be a very long period of time between when the system is developed and when it can actually be used in production. As a result, participants on a specific module may not see the link between development and implementation and so grow restless, causing a potential risk to the project. The linkage between development and "going live" is generally tighter under a phased approach.

What Are the Disadvantages?

Big Bang

Disadvantages of the big bang are essentially the reverse of the advantages of the phased approach:

- huge peak resources may be required;
- fewer resources will be available for a particular module;

- the risk of total system failure may be higher;
- cannot readily go back to legacy system;
- personnel have fewer hands-on opportunities to gain knowledge;
- project managers can't show that it works until the system is entirely installed;
- time between development and implementation may be longer.

Phased

Likewise, disadvantages of the phased approach are the reverse of advantages of big bang:

- heavy use of temporary interfaces;
- need to maintain and revise legacy software;
- higher risk of uninvolved and uncoordinated personnel;
- higher risk of losing personnel to turnover;
- may not be enough modules implemented to achieve functionality;
- operative legacy system constitutes fallback position that may derail new implementation;
- longer duration to install;
- higher total cost.

Organizational Characteristics and Implementation Methodology

A cost–benefit analysis, adjusted for risk concerns, should drive the choice of implementation methodology. However, it is difficult to measure risk and benefits. As a result, organizations make choices of implementation methodologies within the context of other issues, some of which we have just summarized. In addition to issues such as peak resources, functionality, management support, and risk, there are other intervening variables that can drive the decision. In particular, these intervening variables include such organizational characteristics as size, complexity, structure, and controls.

Organizational Size and Complexity

Organizational size and complexity are a critical set of intervening variables upon which choice of appropriate implementation methodology depends. In general, smaller and less complex organizations use big-bang approaches whereas larger, more complex organizations use phased approaches. As the size and complexity of firms increase, the likelihood that those firms will pursue a phased approach increases; this trend is summarized in Figure 11.1.

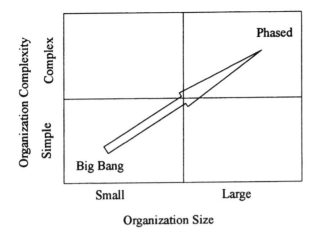

Figure 11.1. Linkages between Organization Size and Complexity
and Implementation Approach Used

Organizational Size and Complexity. Organizational complexity derives
from a number of sources, including the organization's products and customers.
For example, firms with more products or customers are usually more complex.
Complexity can also derive from the characteristics of the customers or products. For example, more powerful customers result in greater complexity than
less powerful customers, since powerful customers can have an impact on the
processes used in the system design. Product line can also drive complexity.
For example, a firm with a broader product line is likely to have a more complex system than a firm with a narrower product line.

The size of a firm also can relate to a number of factors. Although the classic
measure of a firm is revenues or total assets, other factors could include number
of offices or geographic regions. Products and customers might also be used to
measure organization size.

Small, Less Complex Firms. For smaller and less complex firms, there will
be less variation across products and customers. As a result, implementation
of the resulting design is not as difficult as in more complex settings. Further,
since the size is small and the design is not complex, there is likely to be less
risk of failure associated with a big-bang implementation. Accordingly, since
big bang is generally faster and cheaper, smaller and less complex firms can
effectively employ a big-bang approach.

Large, More Complex Firms. If the firm is very large and complex, then a
phased implementation is more likely to be used than big bang. Organizational
size and complexity can make a big-bang approach too difficult or the probability of failure too high, leading the organization to choose a phased approach.
Analysts (such as Bobby Cameron of Forrester Research) have suggested that

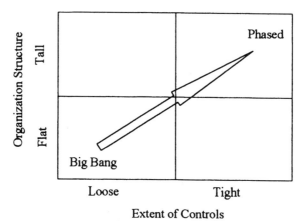

Figure 11.2. Linkages between Organization Hierarchy and
Control and Implementation Approach Used

most large firms use some version of a phased approach and that few large firms
employ a big-bang implementation:

> It's too difficult – especially for large companies – to run a big project ... and the
> risks in terms of project management are huge .... Huge do-it-all-at-once deals
> are quite tough, and most people are looking for smaller piece-by-piece imple-
> mentations. (Radosevich 1997)

In addition, with a large firm the peak resource requirements for implementation could be overwhelming.

Organization Hierarchy and Control

Even if all other organizational issues are equal, the choice of implementation
methodology should consider organizational hierarchy and controls. In general,
as an organization becomes more hierarchical with tighter controls, it becomes
more able to sustain a phased implementation. This relationship is summarized
in Figure 11.2.

Flat Organization and Loose Controls. Bobby Cameron of Forrester Research has suggested that some company characteristics should lead to use of
a big-bang approach: "If a company has a flat organization that is not tightly
controlled, it's very difficult to sustain commitment throughout a phased implementation" (Radosevich 1997).

Extensive Hierarchy and Tight Controls. If the organization structure has
an extensive hierarchy (is "tall") and there is tight control, then there is substantial organizational machinery to facilitate a phased implementation. In addition, because of the organization structure and controls in place, it may be

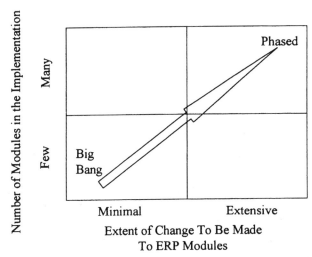

Figure 11.3. Linkages between Implementation Approach and ERP Modules

that – absent substantial reengineering – a phased approach is the only politically feasible one.

Extent of Implementation

The extent of the implementation – as characterized by the number of modules and the degree to which the organization changes those modules – can also influence the implementation methodology, as summarized in Figure 11.3.

Number of Modules

Since ERP systems are modular, organizations can choose to implement different modules that meet their needs. For example, Microsoft installed only the financial module as part of its SAP R/3 implementation. Other organizations implement full suites of ERP systems.

As the number of modules increases, it becomes increasingly difficult to coordinate all the module interactions. In addition, the amount of resources required increases with each module chosen for implementation. As a result, as the number of modules increases there is a shift from a big-bang to a phased approach.

"Module Fit": Extent of Modification

In some cases, modules will require only limited modification, but in others they may require substantial modification. For example, as noted in the case of Siemens (Hirt and Swanson 1998), the financial modules fit nearly 100% of the firm's requirements; other modules fit only 70% to 80% of their needs and thus required some modification.

As the extent of change to modules increases, preferences will shift from a big-bang to a phased implementation. If the modules used are virtually as the vendor intended and developed them, then interaction problems will be minimal.

Few Modules and Minimal Change

If there are minimal changes to the modules then the extent of testing is largely limited to vendor-tested scenarios, ensuring the modules were properly implemented and that best practices interface with each other appropriately. Of course, if there are only a few modules then there will not be as many interactions to be tested. This means that an organization choosing to implement few modules with minimal change is likely to employ a big-bang approach.

Many Modules and Extensive Change

Changes to the software force increases in the level of complexity of the implementation. In particular, software changes introduce the possibility of errors and force the need for increased testing of those changes – within both the specific module and the modules it interacts with. The need to implement a large number of modules generates an even higher level of complexity. Hence, if an organization has chosen to implement many modules and if some of those modules will require substantial change, then a phased approach would be expected; in fact, high levels of complexity may make a phased approach necessary.

Alternative Implementation Approaches

A number of other terms have found their way into the implementation of ERP systems, including "waved" and "aggressive phased" implementations. In addition, terms such as "running in parallel" have been a part of the heritage of computer system implementations.

Waved

Tektronix employed an alternative approach, referred to as *waved* (Westerman et al. 1999). Their ERP implementation was viewed as a change program that consisted of different waves of change. Each wave in the program would deliver functionality to a different business unit or geographical region. Although each wave would be managed independently, the overall project team would manage the interdependencies to ensure that the change program remained on course. For example, Tektronix took three years to implement its Oracle applications.

In the first year, it implemented the general ledger in sixteen countries. Then Tektronix focused on converting the accounts receivable and cost management in its regional accounting centers throughout the world. After the financial applications were implemented, Tektronix was able to add new capabilities (e.g., reports of bookings, billing, and backlogs) as part of their business unit product information.

The wave approach provides a number of advantages. First, each wave supplies feedback as to how the system implementation is proceeding and how well it is being accepted. Second, with each successful wave there is a victory that can keep team morale high. Third, waves are flexible; if a new release of the ERP software becomes available, then the program managers can add a new wave.

Aggressive Implementation Schedule

Although some organizations are reluctant (for any number of reasons) to pursue a big-bang implementation, they may still pursue what is referred to as an aggressive implementation schedule. An aggressive schedule suggests that temporary links to legacy systems really are temporary and not semipermanent. In addition, an aggressive schedule might implement multiple modules at the same time.

Running in Parallel

A classic implementation strategy is to run in parallel. If an ERP system is being implemented then the organization would run the ERP system and the existing system for some period of time (typically, a month or a quarter) in order to ensure that the new ERP system was running as anticipated. This approach can be used regardless of how many ERP modules are implemented.

Running in parallel has a number of advantages. The old system provides a basis of comparison to see if the new system is working as anticipated. Similarly, the old system provides a back-up in case the new system does not pan out. As a result, running in parallel is often seen as a low-risk solution.

Unfortunately, running in parallel has a number of limitations. First, it requires roughly double the number of computing and human resources – a substantial drain for what can be an extended time period. Further, although the concept behind running in parallel is to test the new system against the old system and see if it is correctly capturing the transactions, there are reasons to suggest that the old system may not be a good basis for comparison. Those running the old system have little incentive to keep that system going, since it will be replaced at the end of the test; as a result, the old system may be seeded with substantial errors. Moreover, if the old system is functioning then some may not see the need to proceed with the new system. Accordingly, the continued existence of the legacy system may threaten the potential life of the new system.

Reengineering changes processes and information requirements. The legacy system and the new system may thus have little in common, so the former may not provide a meaningful basis of comparison. Therefore, running in parallel is less likely to be a useful strategy if there has been extensive reengineering.

Summary

This chapter presented a discussion of the advantages and disadvantages of phased and big-bang implementation methodologies. In addition, it also introduced the "waved" approach, which lies somewhere between the other two.

Which approach is optimal, and which approach should a firm use? The contingency models presented in this chapter suggest that there is no generally optimal approach. Instead, firms need to examine such factors as organization size, complexity, control, and hierarchy in order to decide which approach they should use. In addition, firms need to assess the relative advantages and disadvantages before deciding which approach will likely work best for them.

References

Hirt, S., and Swanson, E. B. (1998). "Adopting SAP at Siemens Power Company." Paper presented at the International Conference on Information Systems (Helsinki), December.

Radosevich, L. (1997). "Quantum's Leap." *CIO Magazine,* February 15.

Westerman, G., Cotteleer, M., Austin, R., and Nolan, R. (1999). "Tektronix: Implementing ERP." Report no. 9-699-043, Harvard Business School, Cambridge, MA.

Appendix 11-1

Quantum, I: Requirements, System Choice, and Implementation Approach

Hank Delevati, CIO and vice-president of worldwide information services for Quantum, noted that

This case has been rewritten, reordered, condensed, and integrated based on the following materials: Data Warehousing Institute, "BP Winners," ⟨www.dw-institute.com/⟩, 1998; J. Keerstetter, "Quantum Taps Oracle for Global Inventory System," *PC Week Online,* June 27, 1996; "Oracle at Work with Quantum Corporation," ⟨www.oracle.com⟩ (no date); "1996 Annual Report," ⟨www.uk.oracle.com/⟩, 1996; L. Radosevich, "Quantum's Leap," *CIO Magazine,* February 15, 1997.

[i]t's extremely important for us to be responsive and responsible to our customers. And when it comes to managing our demand and managing our inventory worldwide, [given our information systems, we just could not do it] ....

As a result, Quantum was considering the purchase of an enterprise resource planning system to help meet their needs. Still they needed to understand their requirements, which software met those requirements, and which approach would best facilitate implementation.

Quantum

Quantum, with corporate headquarters in Milpitas, California, was a $4.4 billion company. Quantum was one of the world's largest manufacturers of computer storage devices, including hard drives and tapes. In 1996, Quantum was the second largest producer of server drives, producing over six million. In that same year, Quantum was also the largest selling manufacturer of hard drives for personal computers, selling more than 25 million. Quantum had 23 sites worldwide, with factories, suppliers, and customers spread across North and South America, Asia, and Europe.

Computer storage has been a very competitive and cyclic business. In 1986 there were 350 disk drive manufacturers worldwide. Ten years later, however, there were only four major disk drive firms in the United States, including Seagate, IBM, Western Digital, and Quantum. The market continues to be very competitive; as Delevati noted:

> Disk drives have become a commodity. Although Quantum disk drives are known for their quality, we sell disk drives with reliability, functionality, speed and performance. But customers always want something with more capacity, faster access times and a lower cost. We need to stay on top of that.

Although Seagate and IBM design, manufacture, assemble, distribute, and market their own drives, Western Digital produces their drives in conjunction with other firms. Quantum differed from the other major producers in that they outsourced their production but took responsibility for distribution and marketing.

Quantum's Business Strategy

As disk drives became more and more of a commodity, Quantum's strategy to differentiate themselves was to make customer service their competitive advantage. As noted by Delevati, "[o]ur business model is staying close to our customers. Our customer relationships come first – above the technology, price performance and the form factor."

Quantum's Information Systems

Quantum was limited in its ability to respond to customer needs. Quantum had nine different legacy systems that supported their business units but could not share information. As a result, Quantum's business units were operating semi-autonomously, making it difficult to coordinate the firm's response to customer needs.

In the spring of 1994, Quantum's information system problems were typical of the legacy systems on which they were based. The legacy material requirements planning (MRP) system, AskManMan, kept each division's transactions in separate functional and business unit databases. Information had to be manually consolidated, since databases could not share information. The process of gathering faxes, e-mail, and written information meant that this manual consolidation of MRP data took four days. Closing the books took at least seventeen days. Inventory, accounts payable, and accounts receivable each required consolidation and rolling up for reporting purposes.

Even internally, order information was limited. Inventory availability could not be confirmed and delivery could take days or weeks. In order to process an order, availability information from a number of distribution facilities around the world had to be assembled. This could also take days or weeks. Unfortunately, if delivery took too long, orders could be canceled. As noted by Mark Jackson, an executive vice-president, "our performance ... was below par."

The situation was further exacerbated when, in October 1994, Quantum purchased Digital Equipment Corporation's (DEC) storage business. Digital had a financial system that had been developed in-house as well as a different MRP legacy system, called Maxcim, that was now owned by Computer Associates. In addition, DEC had a number of different legacy systems that could not communicate with each other. As a result, DEC managers and sales people could not get timely, enterprise-wide information. Further, the acquisition doubled the number of system artifacts (e.g., product lists, vendor numbers, and client numbers) at Quantum. Finally, the acquisition also roughly doubled the number of employees and processes used by Quantum.

In order to execute their customer service strategy, Quantum had to enable their sales people to take and confirm an order in real time. In addition, those sales people had to be able to confirm a delivery time and follow up to ensure that the order was delivered – also in real time. These goals required the ability to assess current inventory and to reserve that order for the customer in real time. This ability to confirm and allocate inventory was referred to as "available to promise" (ATP). Legacy systems did not have ATP capability. Unlike legacy systems, an ATP capability cut across multiple locations and multiple modules; ATP was only possible if a sufficient ERP infrastructure was in place.

Quantum's ERP Project

In 1993, Quantum recognized these problems with their systems. As a result, in April 1993, the information systems department began sending out requests for proposals for an enterprise system that would meet their needs. In March 1994, Quantum selected Oracle, Hewlett-Packard, and Price Waterhouse to help them overcome the problems they faced with their existing information systems.

Quantum's Project Team

Quantum's project team had top management support from the beginning. As noted by Jackson, "to succeed you have to spend money and take care of the details." In April 1994, Quantum staffed the project and named it "WARP" (worldwide ask replacement system). Michael Brown, who later became CEO, was put in charge of the project effort. Under Brown's lead, during the summer of 1994, Quantum developed three primary project teams. First, the steering committee was headed by Brown and included vice presidents of finance, information systems, logistics, manufacturing, purchasing, and sales, as well as representatives from Price Waterhouse. Second, there was a core team consisting of sixteen managers from a number of Quantum's departments and business units. Third, a 100-person project team – consisting of members from the information systems department and several of each business unit's key employees – was formed. The project members from the last two groups were removed from their day-to-day responsibilities and put on the project full time. Cross-functional teams were chosen to facilitate a process focus. Group members were moved to a special location for the duration of the project.

Choice of an Enterprise Resource Planning System

After the teams were set up, processes were analyzed to determine how work got done. Using play acting and other approaches, team members analyzed and improved business processes. Based on those new processes, the requirements for the new system were established. After considering a number of different packages, including SAP's R/3, Quantum chose to implement an Oracle 7 database with Oracle's financial and manufacturing modules. Delevati indicated that Oracle applications were chosen because they were the only ones with ATP capabilities.

Setting the Implementation Date

After deciding on Oracle Applications, the teams brought in other consultants from Price Waterhouse and Oracle. A number of pilot projects were begun. Ultimately, the WARP team expected a summer 1995 implementation.

Problems with the Implementation

Almost immediately there were problems. Acquisition of DEC slowed the implementation effort. Acquisition of DEC's hard drive divisions complicated the design and implementation efforts of WARP. According to Delevati, "[t]he complexity and magnitude of the project quadrupled." There were now duplicate systems, processes, and artifacts. In addition, the project team was asked to help integrate the new acquisition into Quantum. Accordingly, for the next three months the WARP project was put on hold while the new acquisition was handled.

In January 1995, the project team was reassembled. The team had to redo its original work to reflect Quantum's enlarged product line with additional artifacts and processes. In order to address this increased scope and the need to redo previous work, the steering committee increased funding.

Phased versus Big-Bang Implementation

A major decision facing management was whether to employ a phased implementation or a big-bang approach. The decision would influence development, testing, and ultimately the success of the implementation.

With a phased approach, a company typically implements a single module at a time. In some cases, a single application is implemented a single location at a time. With this type of approach, temporary interfaces must be established between new ERP modules and existing legacy systems until the ERP system is fully implemented.

In a big-bang approach, the firm implements an entire suite of ERP applications – at all locations at the same time. For Quantum, using a big-bang approach would mean implementing the 17 Oracle modules at 23 sites around the world, virtually all at the same time.

There seemed to be contradictory information as to which approach should be used. On the one hand, Delevati argued that "[t]he phased approach is longer – and I contend riskier, because you won't get everyone involved and coordinated." The concerns with a phased implementation were further elaborated by Bobby Cameron, a senior analyst at Forrester Research: "If a company has a flat organization that is not tightly controlled, it's very difficult to sustain commitment throughout a phased implementation." However, Cameron also indicated most large firms use some version of a phased approach and that few firms employ a big-bang implementation.

> It's too difficult – especially for large companies – to run a big project ... and the risks in terms of project management are huge .... Huge do-it-all-at-once deals are quite tough, and most people are looking for smaller piece-by-piece implementations.

Questions

1. What requirements did Quantum have that were not met by their current systems?
2. How might Quantum choose artifacts and processes for their ERP system, especially now that DEC has doubled Quantum's business?
3. What are some advantages and disadvantages of phased versus big-bang implementation?
4. Which approach, big-bang or phased, should Quantum use? Why?
5. What kind of firms would you expect to use a big-bang approach, and what kind of firms would you expect to use a phased approach?

Appendix 11-2

Quantum, II: Going with Big Bang

Quantum decided to employ a big-bang approach to implement their Oracle application software. As noted by Mark Jackson, an executive vice-president, "[i]f we had taken a phased approach, I would not have gotten that ATP until way down the path. We *had* to do a big bang."

Testing for a Big-Bang Implementation

Systems Validation Test 1

Quantum planned six months of extensive testing as part of their big-bang implementation. Six months of testing translated to about seven or eight months of elapsed time. As observed by Mark Vito, who was involved as a manager from Oracle's manufacturing consulting group, "[p]eople talk about doing things in certain ways, like extensive systems testing, but in reality, they don't do it … [however] that was not true in Quantum's case." By September 1995, each of the major pieces had been tested using a "conference room" pilot, whereby a particular module is tested with a range of transactions to see if it works as anticipated.

In order to run some broader-range tests, Quantum gathered about 100 users at corporate headquarters. This validation test, dubbed "systems validation test 1" (SVT 1), was designed to be two weeks long; it would test system capabilities, including hardware. A large number of transactions would be entered

See footnote on p. 162 for sources.

by the users, simulating how the system would respond to the future demands placed on it.

Unfortunately, this testing barely made it through the first week. The network crashed. Applications did not work as anticipated. Orders did not lead to changes in factory schedules.

Changes Resulting from SVT 1

After these problems with SVT 1, Hank Delevati joined Quantum to oversee the implementation. Delevati had previously overseen a big-bang implementation of Oracle applications at Sun Microsystems.

Delevati indicated that some of the problems were due to general hardware and software issues. As a result, Quantum moved to the latest-version Hewlett-Packard (HP) 9000 servers and HP UX operating system. In addition, Quantum upgraded to the most recent version of the Oracle applications, Oracle 10.4. This new version required a number of brand new set-ups and had hundreds of new relational database tables that required configuration.

In spite of Quantum's extensive testing plan, certain test aspects were overlooked. Although individual project pieces had successfully passed the tests, interdependencies between different pieces had not been tested. As noted by Delevati, "[i]t was a shock to them that although the pieces worked independently, they might not work together."

In order to coordinate movement toward a second test, Delevati developed an organization chart designed to pinpoint problem areas. Each WARP task was listed and an "owner" was identified. Task categories included business functionality, systems performance, data conversion, IS readiness, training, documentation and individual site readiness. Within each task there could be hundreds of subtasks. Each subtask was associated with a color, indicating what stage it had reached, as follows:

- green = "ready";
- yellow = "possible manual intervention";
- red = "not ready yet."

Once colors were placed on subtasks, the teams could focus on those tasks marked red.

Systems Validation Test 2

As each day passed, more tasks went from red to green. By early December, all of the red tasks had been colored either green or yellow. However, the importance of the success of the systems validation test 2 (SVT 2) was clear. As Delevati observed:

> They knew that if a second worldwide system validation test failed, morale and the momentum of the project would sink to an all-time low or might even start to decay. It had to work.

SVT 2 ran from December 11, 1995 through January 8, 1996. Users from around the world flew to corporate headquarters to participate in the system test.

Although the SVT 2 went well, the system test highlighted a few areas that needed further attention. As a result, Quantum scheduled a final test in February, 1996. In this test, only a few tasks previously colored green had to be colored yellow, requiring additional WARP attention. For example, during the test, the network in Penang, Malaysia was close to overloading. Unfortunately, in Malaysia it took 90 days to increase the bandwidth. In order to avoid delays in the implementation, the team shut off access to the Internet for the staff in Penang until the new bandwidth could be implemented. All the bandwidth would be dedicated to the Oracle applications.

Training

Throughout the system implementation and testing, Quantum sponsored substantial user training. Quantum's 24-month implementation schedule included six weeks of worldwide training. Employees were taken off their jobs to train for periods ranging from two to four weeks. Jackson indicated that "[t]he training was taken very seriously, and local managers were not allowed to pull employees from training no matter what business problem arose."

Big-Bang Implementation

By April 26, 1996, most of the WARP problems had been taken care of. At 5:00 P.M. Pacific Standard Time, the conversion began. All records were converted and transferred to the Oracle applications. On May 3, a snapshot of the data was taken and some transactions were run. Shortly thereafter, the 100 project team members logged on to run some sample data. The system worked. It was time for the big bang.

On May 5, Jackson turned the system on for real. Senior management and the WARP team watched as users from around the world logged on. As noted by Delevati, "[i]t was exhilarating, it was exciting, and there was a lot of energy in the air. We had done our homework. We were sure it would work."

By May 6, 1996, Quantum had converted over $1 billion in accounts receivables and $1.6 billion of open purchase orders at the time of the implementation. With over 1,300 users worldwide, Quantum's system had become the largest distributed Oracle application system to be implemented using a big-bang approach.

Delevati remarked that

> [b]usiness as we knew it stopped. For eight calendar days we could not place an order, we could not receive material, we could not ship products. We could not even post cash.

> No business could be transacted that week other than payroll. Everything from inventory to purchasing to scheduling and shipping was down. Then we brought the whole company up with a backlog of work with all sites integrated around the world.

Oracle Application Capabilities

The system ultimately employed nine HP servers in order to support over 40,000 relational database tables. At Quantum, Oracle's applications primarily are used for inventory control and financials. The Oracle installation allowed integration of both inventory availability and delivery information in the same system. As a result, inventory availability and delivery dates could be confirmed in just minutes.

Adopting the system had additional advantages. As noted by Delevati,

> [o]ne of the most evident benefits is that all of our orders, all of our inventory, and all of our financials are consolidated and centralized. This is a huge competitive advantage. Prior to the new system, seventeen days was the fastest they could close the books. The first month [with the new system] ... we closed the books in nine days. And our first quarter took only eleven days.

Delevati also observed that "[s]trategically we can see what's coming, wind down production of some products, ramp up for other products, and be proactive with customers in life cycle planning. It's a big advantage to manipulate the real time data along with the market data."

Costs and Benefits

Mark Jackson estimated that the project was as costly as developing and producing a new product line. Although no return-on-investment (ROI) study had been made to calculate a measure of success, Jackson thought the implementation was successful and commented on why he thought the project worked:

We could have figured how to save 10% of the project's cost, which would be a significant amount of money, but I think that would have raised the risk to an unacceptable level. To succeed, you have to spend the money and take care of the details.

Questions

1. What are some other reasons that Quantum chose a big-bang approach?
2. What factors contributed to Quantum's implementation success?
3. What did you think of the testing approach implemented by Delevati?

Appendix 11-3

Quantum, III: Collaboration and Competition

On June 24, 1996, Oracle's application division announced that "several companies went live with their Oracle Applications implementations during the quarter, including Silicon Graphics, Inc. and Quantum Corporation, both of whom successfully deployed large-scale implementations." In addition, Oracle's application division announced that "among the customers added this quarter included ... Western Digital ...."

Questions

1. What do you think of Western Digital's choice of Oracle Applications? What do you think that Quantum thinks of Western Digital's choice of Oracle Applications?
2. What capabilities do you think Western Digital would choose to implement?

12

After Going Live

After going live with the ERP implementation, it is not time to forget about ERP because it is "done." Instead, at that time the organization enters a stabilization period that typically drags down organizational performance. As a result, firms need to work to mitigate that negative effect. The firm also needs to build an organization to handle the day-to-day issues associated with the ERP system, and management needs to determine what else must be done and whether the implementation matches the system plan. Also, management must address what upgrades, extensions, or linkages can or should be made to the ERP system. Finally, the firm needs to evaluate the success of the project.

The Stabilization Period

After a system goes live there is what has been referred to as a "stabilization period" that typically lasts from three to nine months. During that period, as noted by a director at Benchmarking Partners, "[m]ost companies should expect some dip in their business performance at the time they go live and should expect that they'll need to manage through that dip" (Koch 1999). For example, a recent survey by Deloitte Consulting found that, after their ERP systems went live, one in four firms suffered some drop in performance.

During the stabilization period, all those processes that once were just plans are now being used. New software and processes may be unfamiliar to the users. Hence, there may be problems with the quality and quantity of the work, and consequently the system may not operate as hoped. For example, in a discussion about a system that went live in January 1998, Koch (1999) notes that

> [c]ustomers did not get deliveries, or they got the wrong amount or the wrong products, you name it. The customer service people had so many questions about the new system that the project team spent much of the first few weeks by their side putting out fires .... It wasn't until five months later in May that ... [the

company] smoothed over the problems with its customers. The system wouldn't calm down until September.

As a result, it is critical that users get the appropriate training at the right time. Training too early may not be remembered, but training too late may not be in time.

During the stabilization period, processing and network response times may not be adequate for the implemented system, forcing changes in bandwidth and processing capabilities. For example, when Cisco went live they found that one of their primary problems inhibiting performance was the hardware sizing and architecture (Cotteleer, Austin, and Nolan 1998).

Where does support for these problems come from? Internally, the original project team can evolve to provide support for user needs, assist with additional training, and make necessary system changes. Koch (1999) quotes an ERP team member as follows: "At the beginning of the project I thought I would be able to go back to my group at the end ...." Instead, while doing that job, she became first the unofficial help desk and later a full-time SAP consultant within the company.

ERP Support Organization

When the system goes live the implementation team needs to remain in order to support the users. Typically, that organization is under the control of the vice-president for information systems or the chief information officer. Activities of the support organization can include:

- detecting and responding to system bugs;
- answering user questions (e.g., through a help desk or training);
- making changes in the system parameters as the organization changes (e.g., as the organization evolves there will be a need for changes in the vendor list, the chart of accounts, and other MAPs);
- ensuring that consistent production versions of the software are available;
- managing different ERP input and output capabilities (e.g., responding to changes in reporting needs);
- maintaining and updating the documentation and training materials;
- maintaining and upgrading the software.

Where do members of this support organization come from? Participants in the original ERP project team can be important players in the day-to-day operations. It is therefore important to retain project team members who have the (highly marketable) skills necessary for day-to-day ERP system operation.

Generally, members of the project team have valuable knowledge that needs to be preserved through their retention. Hence, firms need to advise project team members of career opportunities that will allow them to exploit their knowledge within the firm.

A recent report (Stedman 1998) found that different ERP software packages have different costs of running them. A META Group survey of 50 firms found that the total costs of running a system for two years – as a percentage of annual revenues – ranged from 0.4% for Lawson's software to 0.67% for SAP to 1.1% for BAAN. Of course, extrapolation from such studies may be limited by a number of intervening factors, such as firm size. For example, Lawson usually is implemented in smaller firms whereas BAAN has a reputation for being implemented primarily in larger firms.

Determine What Remains to Be Done or Revised

Determining what needs to be done (or redone) regarding the system implementation can be accomplished by evaluating the data conversion, determining what new process bottlenecks have emerged since system implementation (e.g., during the stabilization period), and assessing whether the documentation and training has been sufficient.

Data Conversion

An important task required for the migration from legacy to ERP systems is that of data conversion. Old data needs to be placed into an appropriate format and sequenced for implementation. Changes to the data derive from a number of different sources, such as the existence of new fields and the development of new MAPs (vendor lists, charts of accounts, etc.).

In many cases the quality of the data conversion effort (and resulting problems) can only be observed over time. Cumulative data needs to be examined for "sense." For example, it may appear that inventories are building up, collections are down, and part outages are occurring. However, these changes may actually signal problems arising from inadequate data conversion: wrong part numbers, vendor numbers, or inventory numbers.

Mitigating potential data conversion problems typically requires a number of actions, including the following. First, reconcile the data in the legacy and the ERP data sources where possible, forcing such questions as: Do the inventory numbers match? If not, why not? Are there the same number of products? If not, why not? Are there the same number of vendors (clients, etc.)? If not, why not? Second, perform a mapping from one database to the other to make sure that all data is included. The mapping may reveal new and omitted product numbers as well as new and omitted vendors and clients. Inventories should be equivalent. Without these actions there is no guarantee that the data conversion did not mix different kinds of data or that all data is accounted for.

Process Bottlenecks

No matter how much effort has been made or how good the ERP implementation and planning was, it is still likely that process bottlenecks and non–value-added activities remain. As a result, firms should establish systematic approaches for identifying such bottlenecks and activities.

Where are those bottlenecks likely to occur? First, their location is likely to be different for different firms. Second, the cross-functional nature of ERP process design is likely to generate bottlenecks, particularly where different departments have different resources. Third, for those settings where data gathering has been transferred from accounting to where the data is actually generated, there may be process bottlenecks because of the change in data input. Fourth, linkages to legacy systems and processes may drag down system performance.

How do you locate the bottlenecks? One approach is to listen to the customers and partners. What are they complaining about? Follow those complaints. Unfortunately, the complaint road can take a while to manifest itself. Another approach is to sit back and wait at the help desk for the questions and resource requests to come flooding in. This approach may also be too late.

Two other approaches can be used to detect bottlenecks: an internal ERP data analysis and an organizational analysis. Internal ERP analysis is based on an examination of ERP data; exception reports and transaction data can be analyzed for emerging and repeating problems. Alternatively, going into the organization and talking with those who use the system can turn up issues before they become major problems.

Documentation and Training

System and process knowledge is required in order for the system to work. As a result, adequate training and documentation are required prior to going live. Since virtually all organizations either underestimate or underallocate for the necessary training, this is likely to be a problem area. However, until the system goes live, the adequacy of training and documentation cannot be truly assessed. Thus, when the system goes live, a review of changes in documentation and additional training begins, and resources can be allocated to where they are needed.

Compare the Plan with Reality

Another source of information about what still needs to be done is the comparison between plans and reality. For ERP systems, this comparison has at least three dimensions: systems design and implementation, planned and actual use, and expected versus actual system capabilities. The basic nature of this comparison roughly equates the comparative analysis to an audit, possibly by different personnel or even an outside source.

These comparisons are critical. If there are marked differences between plans and reality for any of the dimensions, engagement success may be compromised. As an example, the ERP decision may have been made assuming certain changes in best practices, which were expected to lead to organizational improvements in efficiency and effectiveness. However, if the ERP was not implemented as planned then those benefits will not be attained. As a result, the basic reason(s) for choosing the ERP may not have been addressed.

Why would there be differences? Perhaps the biggest source of problems are implementation compromises.

Implementation Compromises

Implementation compromises are deviations from a planned implementation – generally to save time or money. Oftentimes, in the rush to have the system installed, a number of compromises are made. After the system is implemented, it is important to examine what needs to be done or redone as a result of these compromises. For example, as noted by the senior manager of corporate accounting systems for Federal Express, "[t]here are timing and scope issues during the project that cause you to say, 'we'll do this piece later'" (Koch 1999).

What is the impact of implementation compromises? The ERP installation saves time and money over what would have been the cost if there had been no compromises. However, because the full scope has not been implemented, the system and processes will not operate entirely as anticipated. Implementation compromises will therefore affect evaluation. In some cases, implementation compromises may ultimately result in engagement compromises, as in those settings where the move to ERP was done in order to change key processes or make key changes and those changes were compromised.

One implementation compromise is the short-term sacrifice of security. For example, password security may at first be implemented at group levels rather than at the individual level. Another frequent area of implementation compromises is data input/output. For example, after Geneva Steel implemented their system, they found that the reporting needs of users were not being met. In order to meet user requirements they installed a data warehouse that was used to generate reports in response to user queries. This shows how it is oftentimes important to revisit implementation compromises to make sure that any outstanding issues are addressed appropriately.

How do firms identify implementation compromises? There are at least three approaches. First, there can be a comparison of the implementation to the plan to ensure that all was done. Second, the firm can set up lines of communication so that they can effectively listen to their own employees and their problems with the system. Third, firms often bring in consultants to review the implementation to detect such compromises, particularly in such areas as security.

System Design and Implementation

The system design is not always what is implemented. Processes and MAPs ultimately implemented may not match the original plan. In order to find differences, planned implementation should be compared to actual. When differences are found between the plan and actual realization, those differences should be investigated to make sure they were authorized.

Compare Planned and Actual Use

The way a system is used may not conform to its planned use. A system audit can be performed in order to find out how the system is actually being used. For example, reports made available to some users may also be available to other users; whether or not that second set of users should receive the reports may be an issue of concern.

Compare System Functioning: Expectations versus Reality

After the system goes live is when firms get a reality check from top management. At this point, the ERP instantiation of the vision is now real enough to be seen and compared to what was expected. For example, as seen in the Geneva Steel case, the SAP implementation was anticipated to make a number of far-reaching changes in the organization. A comparison between key parts of that vision and the ensuing reality are necessary to ensure that the system is working as expected.

Establish Linkages, Upgrades, and Extensions

After stabilization, resources devoted to implementation can now be used to facilitate linkages to other systems and to extensions of the ERP system that have been identified. Throughout this process, the importance of different linkages, upgrades, and extensions should be prioritized so that (a) maximum benefit can be generated from expenditures and (b) priorities are consistent with the firm's overall vision.

Upgrades

In some situations, upgrades to different system versions must be made so that additional features can be implemented. For example, ABB told each of its thousand companies that they needed to implement activity-based costing (ABC), which ABB Industries chose to implement through SAP. However, after implementing version 3.0 of R/3 they found that they needed a later version to be able to fully implement ABC. This firm's ability to implement a necessary feature – activity-based costing – was tied to their implementing a new version of SAP.

Linkages

Although ERP systems are designed to function independently of other systems, they still can be linked to a range of such other systems. In order to ensure that the ERP system goes live at the planned time, those linkages are sometimes not made until after the "go live" date; such linkages can be pursued after the implementation.

Extensions (New Features and Functions)

Enterprise resource planning systems have become the information system "backbone" in many firms, providing processing capabilities for virtually all of a firm's transactions. Over time, ERP systems have been extended in a number of different directions. For example, ERP vendors have moved to provide supply chain integration and sales force automation support. In some cases, these extensions were known of when the software was chosen and may even have influenced the final choice of software. In other cases there may be new extensions to the ERP software that the firm needs to go back and examine in order to determine whether or not the options should be considered and implemented.

Evaluate Success

After the stabilization period, the firm can begin the process of evaluating success. This requires determining just when the evaluation should be made and exactly how success is to be measured – for example, using cost–benefit analysis, some approach based on the criteria used for deciding to go ERP, or a "balanced scorecard" approach.

Timing

Timing is a critical issue when evaluating a new system. Rarely would benefits be fully realized immediately after implementation, nor would much

Table 12.1. *Actual versus Anticipated*
Project Factors

	As a percentage of expectations, what was the actual:		
	Duration	**Cost**	**Benefit**
<50%	0.0	0.0	6.0
50%–100%	25.0	13.5	65.0
100%–125%	27.5	43.0	8.5
125%–150%	25.0	24.5	14.5
150%–175%	3.0	8.0	0.0
175%–200%	16.5	2.5	3.0
>200%	3.0	8.0	3.0

Note: Results are the averaged percentages of 1998 and 1999 data discussed in Austin and Cotteleer (1999). In some cases, rounding differences prevent columns from adding to 100.

information be available. As a result, evaluation needs to be scheduled at a time when benefits could be realized and measurable. Accordingly, firms typically evaluate the ERP implementation during the first or second year *after* the system goes live. At that point, many system bugs will be ironed out. Further, the system will have generated a year of data that can be used as part of the basis of evaluation – for example, did inventory turnover improve?

Determine Duration, Costs, and Benefits

Project management requires monitoring the project duration, costs, and benefits of the ERP implementation (i.e., "business benefits"). There have been a number of reports of problems with implementing ERP systems in a timely and cost-effective manner. A recent survey (Austin and Cotteleer 1999), summarized in Table 12.1, found that ERP project benefits were more often short of expectations while costs more often exceeded expectations. A possible explanation for this finding is that costs may have been systematically underestimated and benefits overestimated in order to sell the system to top management.

Although project duration was sometimes shorter than expected, more often it was longer; in some cases, duration was almost double what was expected. As a result, it is probably not surprising that ERP implementations have a reputation for being over budget and late (see e.g. March and Garvin 1996).

However, perhaps the most interesting aspect of the survey was that only 44% of the firms formally tracked the benefits. Other than poor project management, there may be a number of reasons for this finding. First, it is generally accepted that it is more difficult to track benefits than costs because benefits are often less tangible. Second, there may be no way to track benefits – particularly if the project was justified via a technological, strategic, or competitive rationale and hence no specific benefits were isolated. Third, firms may view the implementation as a "sunk" cost and thus may see no advantage to assessing benefits.

Choice Rationale

Another approach to evaluation is by determining if the system meets the criteria set out for it in the beginning. In this case, choice rationales (discussed previously) can be used to evaluate success. However, some rationales are more likely than others to facilitate this evaluation process.

Technological choice rationales – such as the need for client server software, solving the Y2K problem, or meeting competition – offer little evaluation specificity. Once the ERP software is decided on, the Y2K and client server issue are satisfied so there is little left to evaluate. Furthermore, a rationale of simply meeting competition may be too amorphous. At one level, simply deciding to install the same ERP software as the competition satisfies the requirements of "meeting the competition." It may thus be crucial to gather more detailed productivity metrics, such as what it means to "meet" the competition, in order to incorporate greater specificity and feedback into the design.

The strategy and business process rationales can provide greater specificity and so facilitate measurement. For example, getting better control of inventory may have been a rationale. Associated with improved inventory control can be specific measures of raw materials, work in process, and finished goods. Levels of inventory, inventory as a percentage of sales, or inventory turnover can be used as specific goals measuring improvement. As another example, transportation and logistics improvement may have been a rationale. Associated with improved transportation and logistics can be specific measures that relate to warehouse costs and transportation expenses. For example, the expected improvement in warehouse expenses due to ERP implementation can be assessed as a percentage of sales.

Specific measures allow detailed examination in the evaluation process. If the measures reveal less than the expected improvement then the firm needs to ask why and to determine if further improvement is attainable. The firm also can address how the ERP implementation might be fine-tuned to allow achievement of specific goals.

Independent Measures or a Weighted Portfolio of Measures

Any evaluation process of an ERP implementation should consider whether individual measures or instead a portfolio of measures should be used to evaluate the project. If single measures are used and all are attained, there is no question as to the outcome of the evaluation. However, if one or more goals are not attained then there is some question as to the outcome of the evaluation. Does one substandard measure mean that the implementation was unsuccessful?

Evaluation problems do not go away when multiple measures are used as a "weighted portfolio." In such a setting, relative weights on the individual items may be needed to establish the overall performance. One emerging approach is that of the balanced scorecard.

Balanced Scorecard

This approach for evaluating performance in organizations was promulgated by Kaplan and Norton (1992) and Kaplan and Atkinson (1998). The balanced scorecard communicates a firm's multiple objectives from multiple perspectives, including financial, customer, learning and growth, and internal business processes. Rather than depending on a single measure or a single type of measure, multiple measures are used as part of the evaluation process.

The multiple perspective and multiple measures approach can be translated to ERP implementation. A balanced scorecard for ERP systems could employ each of the four basic business case rationales (technology, business process, strategic, and competitive) used for deciding whether to go ERP and so generate a portfolio of perspectives and measures that would depend on the particular company and its reasons for adopting ERP.

Budget for After Going Live

Finally, in order to pursue each of the activities analyzed in this chapter, there must be a budget and a corresponding plan to support the complete project. Without the budget, there can be no people and no progress. Complete project management means managing through the entire project life cycle, including after going live. There have been numerous reports of firms that had no budget left by the time the firm got to the stabilization period. Other reports have indicated that firms did not have sufficient budget to include project team members in such efforts and so lost those project members to other ERP installations.

Summary

After the ERP system goes live, there is still much to do. The implementation team must shepherd the firm through the stabilization period. Further, an organization is needed to run the ERP system on a day-to-day basis. After the system goes live, the firm needs to determine what needs to be done and redone. For example, data conversion, implementation compromises, process bottlenecks, and documentation all need to be evaluated to ensure that they continue to meet needs. Firms need to compare their plans to what actually happened in order to determine the extent of implementation compromises. Firms must also look to the future in terms of what upgrades they should pursue, what linkages need to be made to legacy systems, and what extensions (if any) should be made to the ERP system.

Firms need to evaluate the success of the implementation. However, as found in one survey, only 44% of the firms actually do a benefit analysis. Failing to perform benefit analysis may be a function of poor project management, the difficulty of evaluating benefits, the rationale used to choose or evaluate the ERP implementation, or the view that the project is now a sunk cost. In those cases where the benefits are analyzed, cost–benefit duration analysis can be used to provide insights into the success of the engagement. Other means of evaluation, based perhaps on the original ERP choice rationales, could also be used. In any case, because benefits need time to manifest themselves and be measurable, evaluation measurement should not begin until a year after the go live date.

Finally, many firms do not provide sufficient budget for the post-implementation tasks discussed in this chapter. In order to ensure project management over the complete life cycle, an adequate budget is required to ensure resources throughout the entire project.

References

Austin, R., and Cotteleer, M. (1999). "Current Issues in IT: Enterprise Resource Planning." Unpublished presentation, October.

Cotteleer, M., Austin, R., and Nolan, R. (1998). "Cisco Systems, Inc.: Implementing ERP." Report no. 9-699-022, Harvard Business School, Cambridge, MA.

Kaplan, R., and Atkinson, A. (1998). *Advanced Management Accounting.* Englewood Cliffs, NJ: Prentice-Hall.

Kaplan, R., and Norton, D. (1992). "The Balanced Scorecard: Measures That Drive Performance." *Harvard Business Review,* January/February.

Koch, C. (1999). "The Most Important Team in History." *CIO Magazine,* October 15.

March, A., and Garvin, D. (1996). "SAP America." Report no. 9-397-057, Harvard Business School, Cambridge, MA.

Stedman, C. (1998). "ERP User Interfaces Drive Workers Nuts." *Computerworld,* November 2, pp. 1, 24.

Appendix 12-1

A Case Study of XYZ Company:
How Should We Evaluate the ERP Project?

The following ERP metrics have been generated and circulated at our firm. These metrics were put forward as the desired state roughly one year after the ERP system went live.

Inventories

- Reduce raw materials on hand to less than 20 days.
- Reduce work in process by 40%.
- Improve finished goods turnover from 9 to 12 times per year.

General Administration

- Publish financials within 6 business days (currently it is 24).
- Improve customer service levels to 99%, with 90% on-time delivery (currently it is 97%, with 70% on-time shipments).
- Improve productivity of manufacturing, sales, finance, purchasing, engineering, and information systems by 10%.

Division A

- Improve gross margin by 6% of sales.

Division B

- Improve gross margin by 2% of sales.

I am concerned about them since it is not clear to me if they are metrics or goals. In addition, the implementation does not include any functionality that is

---

The company name is fictitious, and some of the numbers have been camouflaged. The enquiry is from a manager who reported directly to a vice-president in this medium-sized firm.

not already nominally present. As a result, I would appreciate your comments on the following questions.

(1) What are your thoughts on these metrics?
(2) Are they metrics or are they goals?
(3) If they are goals, are they appropriate for an ERP project?
(4) If they are metrics, do they serve to measure factors attributable to the ERP implementation itself?
(5) Whether they are metrics or goals, do these points individually or collectively represent a reasonable expectation for the management team regarding their ERP project? If not, what kind of metrics or goals would be appropriate?

Appendix 12-2

Deloitte Consulting Post-Implementation Checklist

Enterprise resource planning projects succeed best when you know what you're going to do after you get the software up and running. Ask yourself these questions concerning your post-implementation plans to see if you and your team are prepared. If you answer No to any of them, take steps to turn that No into a Yes as soon as possible.

The Basics

Do you have a post-implementation plan?

Benefit Targets

- Have you made ERP business benefits and capabilities part of the business plans of specific general managers?
- Have the projected business benefits of the ERP software been communicated to the rest of the organization?

Project Teams

- Is your project team still in place?
- Has the organization developed a retention plan for "hot skills" employees?

This checklist was given in C. Koch, "The Most Important Team in History," *CIO Magazine* (October 15, 1999), where it was attributed to Deloitte Consulting.

- Are project team members aware of internal career opportunities available to them?

Business Metrics

- Are business metrics in place to measure the project's intended benefits versus what has actually been achieved?
- Has an owner been assigned to track each metric?

Project Management

- Have programs been put in place to help individuals cope with major role changes and the stress that comes with them?
- Has a center of expertise or ERP support organization been established to handle user questions, maintenance issues, and upgrades?

Process Expertise

- Is management more process-oriented than it was before the ERP effort?
- Is there a reward program for employees who use the system?
- Are requests for new reports reviewed and approved by a process owner (to avoid clogging the system)?

ERP Foundation

- Are there plans to extend the ERP system's functionality?

Training

- Is the training of your users job-based (beyond simply learning how to use the software)?

13

Training

One of the most important issues in the ERP engagement is training. According to the CIO of Brother Industries, "[t]he easiest mistake to make is underestimating the time and cost of training end users" (*Fortune* 1998, p. 151). An implementation will be a failure if the software runs perfectly, but employees don't know how to use it. Despite the importance associated with the need for ERP training, a Benchmarking Partners survey of 150 sites found that 43% indicated that the amount of training was the biggest surprise encountered (Stedman 1998b).

However, training encompasses more than just users; the training concerns of implementors must also be addressed. Moreover, training is not simply one step in the life cycle just before or just after implementation. Rather, training is embedded in each part of the life cycle. We therefore discuss training within its own chapter.

Frequently Asked Questions about User Training

- How should user training be timed?
- How much training should users get?
- How do you make up time lost on training?
- When should you train?
- How can employees be encouraged to train?
- How much should training cost?
- What is in the training materials: information technology or rather business information?
- Should you use script-based or general support materials?
- How should training be structured?
- Can faster training be developed?

How Should User Training Be Timed?

Training that occurs too far in advance of the go live date will likely be forgotten. Training that occurs too late will not be done in time and can lengthen the stabilization period. In order to fit training between the current time and the go live date, the firm must consider the amount of time in that gap and the amount of time required by the users to learn. For example, Purina Mills started to train users four months in advance on SAP's R/3 (Stedman 1998a).

How Much Training Should Users Receive?

The amount of training required is a function of the particular module for which users are being trained. As noted by Stedman (1998a), in some cases it can take up to six months for users to get comfortable and proficient with the ERP software. Stedman (1998b) reports that a group of finance workers trained for seven hours daily during the last month before the SAP R/3 system went live at Purina Mills.

How Do You Make Up Time Lost on Training?

Time spent on training is time not spent on day-to-day activities. Not surprisingly, there have been a number of different solutions used to ensure that workers get enough time off for training. At Purina Mills (Stedman 1998b), managers put in extra hours in order to accommodate training hours. At Microsoft (Bashein, Markus, and Finley 1997, p. 71), users were expected to do both jobs by putting in extra hours. Still other firms have made use of temporary employees.

When Should You Train?

The hours assigned to training signal how important it is to the implementation. Training scheduled during working hours indicates its importance, whereas training scheduled outside of working hours suggests that training is not as important as day-to-day responsibilities.

For example, one plant CIO indicated that corporate was planning on requesting 80 hours of education for employees. The CIO felt that the 80 hours of education would not be necessary for his employees and that the training could not be done during normal working hours. As a result, he was planning for his employees to come in and train on three successive weekends. Such a strategy leads to at least two questions.

(1) *How will the employees see the relative importance of the training?* Employees will probably perceive the training as less important than their

Table 13.1. *Actual versus Anticipated User Training Costs*

How close was the cost of user training and other ERP deployment expenses to your original estimate?	
More than 50% above expectations	8%
10% to 50% above expectations	26%
Within 10% of expectations	58%
More than 10% below expectations	6%
Don't know	2%

Note: Based on 50 large U.S. companies surveyed in August 1998.
Source: Forrester Research, Inc. (Cambridge, Massachusetts).

day-to-day work. As a result, they may give training less attention than is necessary.

(2) *How much do you think the employees will take away from this kind of weekend training?* Employees will probably not learn as much in this weekend environment as compared with training received during working hours. Many employees are likely to miss sessions; with one missed session, the rest can become a blur.

How Can Employees Be Encouraged to Train?

The most pressing activities often get the most attention – to the detriment of other, less pressing activities. This means that productive work usually receives greater attention than training activities, and so user ERP training might be pushed aside if firms do not ensure that users take it seriously. As a result, firms have introduced different penalties and incentives. During Microsoft's SAP implementation, for example, training was deemed mandatory for certain critical users. Users that did not attend training were threatened with having their computer accounts turned off. As noted by a senior accounting manager, "[w]e only had to turn off 20 or so accounts" (Bashein et al. 1997, p. 71).

How Much Should Training Cost?

Training costs will vary across ERP implementations. However, as noted by Stedman (1998a), the training budget can be 10% or more of the total project budget. Many firms experience budget overruns in training. In a recent survey of 50 firms (see Table 13.1), 17 indicated that training-cost overruns were 10% or more of the original budget.

What Is in the Training Materials: Information Technology or Business Information?

Training users on how to use an ERP system is a mix of technology, processes, and domain-area content in order to provide a context for the system. Otherwise, training can degenerate into the "ERP Hokey Pokey," where users are advised: "you put your right hand on the …, then hit …." For example, a senior accounting manager remarked of the Microsoft SAP implementation that "I've taught a six-hour course on entering journal entries into SAP 20 times now. It was as much a general ledger course as a system tool course" (Bashein et al. 1997, p. 71)

On the other hand, user training can become too general and not aimed at the specific software. For example as SAP went from R/3 version 2.2 to version 3.0, they tried to adjust their training to match the functionality of the software, but the new functionality was not clear. Thus, as noted by Hirt (1999, p. 13) in her discussion of the Siemens SAP implementation, one participant remarked: "the training offered was 'core' – in other words, useless."

Script-Based versus General Support Materials

There are at least two different kinds of support material: general support and script-based. *General support* is typically context-independent, listing capabilities and defining terms, in a manner analogous to "help" documentation.

Siemens Power Company built *script-based* user manuals, which are designed to take an end user step-by-step through an entire transaction (Hirt 1999). These user manuals were then made available on the local area network. The script-based approach has a number of advantages. First, the materials provide a place for users to find the answers to frequently asked questions. Second, they also serve as documentation of how to use the system. Finally, such materials are context-based, providing the user not only with information but also with context.

There are, however, some disadvantages to script-based materials, as observed by one end user at Siemens Power:

> We can follow the scripts that they gave us, but what if something happens that is not on those scripts? We are finding that we don't know what to do. Then we are kind of fumbling around in the system …. And we get this panicky feeling because we don't know how to look it up. (Hirt 1999)

How Should Training Be Structured?

Generally, most training programs come from an analysis of what the specific company has and what they need to accomplish. Almost all ERP approaches to training have an element of classroom training. However, other formats used include training over the Internet, computer-based training, and self-study.

Based on conversations with consultants, one approach that is consistently well accepted involves designating a member (or group of members) of the client organization as "super users," who then can be responsible for training others. Training super users has a number of advantages. First, this approach has been found to facilitate buy-in from the users, because those who supply the training are people the users know. Second, the existence of super users shows the other users that learning about the system is important. Third, developing super users develops an important understanding at the user level.

Can Faster Training Be Developed?

Because ERP engagements are often behind schedule, firms may try to speed their training. For example, in the Microsoft SAP engagement, "the SAP project team redesigned SAP's training to be as short and fast as possible" (Bashein et al. 1997, p. 71). The ability of firms to speed up training depends on the firm's needs, personnel, and previous training. Some personnel are likely to be quicker learners or able to spend more time on training than other personnel. To the extent the new system is similar to the old system, training will likely be faster. However, trying to speed training is potentially dangerous, with a high cost of failure.

Training the Implementor

System users are not the only ones who must receive training on the ERP system. In addition, those in the firm responsible for implementation also must receive appropriate training.

The specificity of the training on the particular ERP software version can be critical to rapid and effective implementation. Inadequate training can negatively affect the implementation. For example, not knowing what the system is capable of can lead to unnecessary changes, slowing the engagement and increasing the cost (cf. the Dow Corning ERP engagement as described by Ross 1998).

In early 1995, Dow Corning's senior management agreed to implement SAP's R/3 as part of a firm-wide reengineering program. In June 1995, 40 of the firm's top personnel were chosen for an implementation team. Since Dow Corning planned to support SAP in-house, they decided not to use consultants. As a result, it was critical to have the project members well-trained. In September 1995, that training began. Unfortunately, as noted by Ross (1998, p. 14), although "most of the team members attended an extensive training session offered by SAP in September 1995, much of the training was on SAP R/2 and was not directly transferable to R/3." As a result, after training, team members still did not understand how to design processes and configure the software. This inadequate training probably delayed the first pilot implementation, completed in March 1997.

The lack of adequate training also influenced the amount of and quality of the changes made in the ERP system as part of the implementation. For example, as noted by one member of the project team:

> In accounting we decided to value all our transactions in a certain way for management and do it a different way for statutory requirements, and SAP to our knowledge didn't handle that requirement. So we had to put some programming in to convert one to the other. Well now that we've completely examined our process and we understand what it is we really need to do and what SAP can really do for us, we'll end up with a far superior process and not have to use the changes that we made in SAP. (Ross 1998, p. 14)

Summary

Training is an issue of concern in each portion of the life cycle. This chapter focused on training issues for users and implementors. A number of frequently asked questions for user training were addressed. In addition, a case was used to illustrate some of the problems that result from inadequate training for the implementors.

References

Bashein, B., Markus, L., and Finley, J. (1997). *Safety Nets: Secrets of Effective Information Technology Controls.* Morristown, NJ: Financial Executives Research Foundation.

Fortune (1998). "Managing ERP." February 2, pp. 150–1.

Hirt, S. (1999). "Siemens Power Company: Surviving R/3." Unpublished case study.

Ross, J. (1998). "Dow Corning Corporation." Paper presented at the International Conference on Information Systems (Helsinki), December.

Stedman, C. (1998a). "ERP User Interfaces Drive Workers Nuts." *Computerworld,* November 2, pp. 1, 24.

Stedman, C. (1998b). "Surveys Show a Training Surprise." *Computerworld,* November 2, p. 24.

PART FOUR

ELECTRONIC COMMERCE AND RISK

14

ERP and Electronic Commerce

Electronic commerce has emerged as a critical aspect of doing business. For many firms there is no separating electronic commerce from regular commerce; the two have become increasingly intertwined. For Cisco, orders over the web have increased from less than 10% in 1996 to roughly 85% in 1999. As a result of these dramatic shifts, there has been an equally dramatic set of changes in the way business is done using electronic commerce. Increasingly, firms such as Cisco are using electronic commerce to simplify their business processes and improve customer relationships (Riggs 1999).

The electronic commerce phenomenon has occurred at roughly the same time as the ERP phenomenon. Further, both are highly dependent on changes in processes to accommodate the new technology, which leads us to ask: "What are some of the relationships between the two?" This chapter explores that question by first establishing ERP as one of the primary building blocks of electronic commerce and then eliciting the role of ERP in each of a number of major electronic commerce initiatives: order tracing, order configuration, order placing, vendor-managed inventories, build to order, and merge in transit.

Building Blocks for Electronic Commerce

What are some of the building blocks for electronic commerce? ERP provides a repository for the information. Communication networks facilitate access to the information. Wide availability and easy-to-use access make it possible to get ERP information over the Internet.

ERP Is a Building Block of Electronic Commerce

Recently, the director of E-business applications at 3Com commented, "ERP is a building block of E-business" (Ricadela 1999). Why?

195

The ERP system provides the central clearing house of real-time information. For example, at Compaq, virtually the entire base of global daily transactions are available in their ERP system. As noted by a Compaq vice-president,

> if we sell a product in Switzerland today, we are able to record its identity, the revenue attained and the cost of that product. Our IT capability could even make it possible to track that information on a daily P&L [profit and loss statement]. (Teresko 1998)

The ERP system provides the current inventory and pricing information, so that firms know what they have available to sell. The ERP system also has information that relates to product configuration, which is necessary for materials requirements planning. Thus, given information that x final products are necessary, all the inventory necessary to build those x products can be generated. The ERP system provides much of the infrastructure for electronic commerce by providing basic information of this nature. As a result, the ERP systems can be at the center of the electronic commerce world.

Communication between Enterprises for Electronic Commerce

The source of an enterprise's information is the ERP system. However, there is still the need for information exchange between different enterprises. Historically, EDI (electronic data interchange) has been used as the basis for electronic communication of much of that information. Increasingly, web forms technology (WFT) is being used as another way of exchanging transaction information. Typically, EDI is used when there are many transactions whereas WFT is used when there are only a few transactions being originated by any one customer.

The EDI is based on exchange of information in standard chunks (e.g., "invoices") in a standard format (e.g., standard sequence of information). Traditionally, EDI has employed value-added networks (VANs) that provide security and infrastructure for transmission of transaction information. Using WFT, a customer provides information on a form that is available on the Internet or on an intranet or extranet. That information may or may not be directly interfaced with a database or other application.

How important is EDI to firms implementing ERP? In a Deloitte Consulting survey, EDI had the largest number of respondents to the inquiry regarding which additional solutions were implemented to complement an ERP system (Kersnar 1999).

What is the role of EDI and WFT in electronic commerce? Most current-generation ERP systems accommodate EDI, which still plays an important role in transaction processing even for high-technology firms. For example, as noted by a vice-president at Compaq, "EDI brings in orders from some of the [larger] channel partners" (Teresko 1998). The EDI protocol is quite robust and has found extensive use in such electronic commerce–based initiatives as merge in transit (to be discussed shortly). The use of WFT has exploded. For example, roughly 85% of Cisco's orders are reportedly placed on the company's website (Riggs 1999).

However, some difficulties are emerging with traditional EDI. First, EDI is standardized, so in many cases EDI forces use of particular processes and artifacts that may not be optimal. As a result, changes in processes and artifacts over time are limited by the structure of EDI. For example, as noted by a vice-president at Compaq, "[s]ince EDI is really a serial process and basically dictates certain times that you do things, we will eventually have to migrate to an Internet/web-based type of transaction with our suppliers and our customers" (Teresko 1998).

Second, EDI is primarily for larger firms and larger transaction volumes. As noted by one of Dell's vice-presidents,

> [l]arge-scale business-to-business customers will integrate their purchasing and ERP systems with our systems, so it will be completely computer-to-computer.
>
> Premier Pages present the business customer with a password-protected view of the specific pricing they negotiated with us and the status of their order. It's a personalized view for the business customer into the Dell website .... Think of Premier Pages as being in the middle between the public website and the truly integrated ERP solution. The large customers will use the truly integrated ERP solution, using Premier Pages primarily for information. Midsize and small businesses won't have elaborate hookups and ERP systems, so they'll use Premier Pages as one of their predominant methods. (Wagner 1999)

As a result of the size limitation on firms, new opportunities are being developed that integrate EDI and web forms technology. For example, Cisco has its smaller suppliers use a web service provider for business-to-business communication with Cisco (Karpinski 1999). A third problem with traditional EDI is its cost. Value-added networks are much more expensive than using the Internet, so increasingly firms are using EDI over the Internet.

There are also limitations of WFT, which is primarily useful in those settings where the originator has only a few orders but the receiving firm has many. If the originator firm has many transactions then the WFT ordering process may be too costly and time-consuming. Not only must the order be made using WFT to the vendor by the customer, but then the originator must update its own system. Hence, using WFT will cause double the data entry time and thus require more personnel.

Widely Available and Easy-to-Use User Access

The other building block is user access. Reportedly, many ERP systems are accessible via the Internet, and users can access an ERP system over the Internet to place orders. For example, as noted by Trommer (1998), Fujitsu PC Corporation uses its ERP and other configuration software to allow the user (either a Fujitsu sales representative, a reseller, or an end user) to order over the web.

Unfortunately, there are limitations to this approach. As noted by a practice manager at Cambridge Partners (Kersnar 1999), "at the moment you need to be in the office to use ERP." Moreover, the available modem speed may be too slow for effective work.

Similarly, there can be standard requirements for client computers in order to facilitate and enable client–server interaction. However, in the Internet scenario, users' computing environments can be quite heterogeneous. As a result, one approach has been to provide a different kind of client server access whereby the user needs only a browser and Internet access. In this emerging environment, the user has Internet access to a time-shared client that interacts directly with a server in an appropriate standard environment. Using this approach eliminates much of the standardization that can be required at the user level in client server computing, and it also extends the classic client server paradigm to a time-sharing model that is analogous to mainframe computing. In the appendix to this chapter we offer a more detailed example of how this emerging version of client server can be used to facilitate Internet access to ERP.

In addition, as noted in Chapter 4, ERP vendors have developed ERP portals to facilitate access to their systems. The common denominator of those portals is that they are based on web browsers, improving ease of use of ERP systems through web access. Figure 14.1 shows the J.D. Edwards ActiveEra portal, which provides web access to their ERP system.

Enterprise resource planning portals enable the firm to personalize access to ERP systems in a web environment. In addition, portals provide access to a number of related services, thus facilitating further electronic commerce. For example, at SAP portals there is access to information about additional SAP (and related vendor) products and services.

ERP and Customer Orders of Goods

Facilitating the ordering of goods is perhaps the most critical task in commerce. Unfortunately, ordering goods is a process that often is filled with errors and resulting ill-will. For example, Cisco found that 25% (Roberts 1998) to 33%

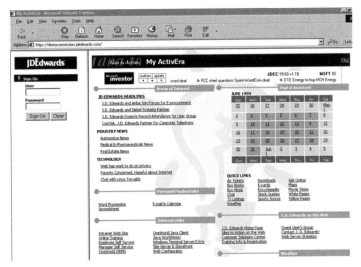

Figure 14.1

(Messmer 1999) of the orders made by faxes had errors in them. Such errors can delay shipment or cause a change in the price. As customers found this out they were forced to contact Cisco about their orders. These requests in turn required increases in Cisco's personnel for responding to customer inquiries, raising costs and slowing down the process of delivering goods to the customer. As a result, Cisco turned to electronic commerce in order to facilitate the ordering process.

Customers Trace Availability of Goods

In 1996, Cisco built one of their first electronic commerce applications (Roberts 1998). Using the web, customers were able to gather information from Cisco's ERP system that would allow them to track and price their orders. For example, they were given access to pricing, configuration, and order status – information that was available seven days a week, 24 hours a day. This strategy reduced Cisco's need for personnel to track orders and answer customer questions. Customer support had been shifted to the customer.

Customers Order over the Internet

Allowing customers to track their orders and find errors was only part of the solution at Cisco. Perhaps a more important question was: "How can you eliminate errors to begin with?" In Cisco's second year, their goal was to eliminate the errors and allow the customer "anytime, anywhere" ordering over the Internet (Roberts 1998). By accessing information from the ERP, customers were permitted to originate, configure, price, and place their orders.

To encourage customers to use the ordering process available on the web, Cisco guaranteed that pricing and configuration would be accurate if the customer used the web application. Within four months, 10% of the orders were being received via the Internet; by 1999, 85% of the orders came in over the Internet (Riggs 1999).

The advantage of this approach is that customers can place highly accurate orders, and within fifteen minutes or less the order will be in Cisco's ERP system ready to be tracked by the customer (Messmer 1999). In addition, within that same fifteen minutes, Cisco can begin manufacturing the goods (Riggs 1999).

The system's accuracy derives from the "configuration engine" (CE). The CE examines orders to find common errors. If errors are found then the engine won't let the customer make the order. The CE examines all available account information and purchase information in order to detect, for example, incorrect part numbers. However, there are still limitations to this approach. For example, the purchasing is done using Cisco's system and not the customer's system. Further, the customer needs to enter the same information into their own ERP system.

Downloading Information to Purchasing

In 1998, Cisco began working with their biggest customers to integrate order information into those customers' purchasing systems (e.g., their ERP systems). New configuration, order, and pricing information is made available once a day to those special customers. Now, those customers can place orders from the familiar systems that they use every day; there is no longer any need to duplicate the process by also entering the order at Cisco's website.

What Is the Downside to Allowing Access to ERP Information?

An AMR research study (Booker 1999) found that, even in the high-tech industry, firms are unwilling to allow access to order information. Although 97% of those surveyed have a website, only 21% of these firms allow on-line access to account information.

What is the downside to allowing direct access to information? First, there are security concerns that include protection of the ERP system, the information, and transmission of data over the Internet. As a result, EDI VANs will continue to be used in some settings. Second, firms may be concerned that customers have access to a "window" on the firm's processes. For example, customers can "see" how responsive the company is to orders that are placed. Dates that goods were produced or transactions were originated can be traced. Order tracking access can provide insight into how much time is actually spent producing a product or how long after the order was placed that production began. Third, access to ERP-based information would need to be accounted for in the capabilities of the hardware and software. A large number of Internet queries could drive down ERP performance for internal availability.

Why Is It Important to Allow Access to ERP Information?

Given these concerns, why does it remain important to provide access to accurate and real-time ERP-based information? First, access to ERP information can improve accuracy, as seen in our discussion about Cisco's implementation. Access to ERP information ensures that ordered items actually are in inventory. Second, access to ERP information can help generate improved business relationships, according to 84% of information technology managers in a recent survey (Booker 1999). Customers can check their account and answer their questions whenever they want to. Third, as noted by the president of Interlink Communications, "I like to use systems for the transaction-oriented stuff and save people for the relationships and support" (Booker 1999). Interlink has opened up information on their Great Plains financial system so that customers can find their account status and detailed invoice information. The incremental cost of any query is virtually zero, and personnel that used to perform that function have been shifted to other customer support activities.

ERP and Vendor-Managed Inventories

Under Cisco's model, the customer does the ordering. In many settings, however, the order process has been shifted from the customer to the vendor. For example, Procter & Gamble (P&G) monitored demand and took responsibility for keeping its products on the shelves (McKenney and Clark 1995). This has a number of benefits for both parties. The vendor has greater incentives to do a better job of keeping its products on the shelves, increasing its sales. Further, if the vendor is managing inventory then fluctuations caused by erratic or inconsistent order processes can be reduced. By improving the consistency of keeping its product on the shelves, the vendor can smooth demand and thus better control productive, transportation, and administrative resources and hence costs. The customer gains because some of the ordering function has been outsourced, reducing its costs as well. Further, if the vendor is able to do a better job of keeping product on the shelves, then there should be greater sales. Overall, costs should go down and revenues should go up.

Vendor-managed inventories are accomplished by providing vendors with real-time access to necessary information. Access must be electronic and the information must be up-to-date or else the quality of the inventory decisions will be compromised – a particularly important concern when the vendor is managing the inventory. In an ERP-based world there are two solutions designed to facilitate vendor-managed inventories: integrating through ordering data, and direct ERP-to-ERP connection.

Integrating via Common Data

One approach to facilitating vendor-managed inventories is to provide access to the necessary data (e.g., sales information) to the vendor. Providing the vendor direct access to a data warehouse with the necessary information is one approach. As noted by a Compaq vice-president,

> [w]e've also custom-developed some tools that sit on top of the SAP ... system to give us a data warehouse capability .... We developed an EDI capability that feeds into our data warehouse. Every week our suppliers use EDI to report on their delivery capability and status. (Teresko 1998)

Directly Integrating ERP Systems

Increasingly, there are reports of directly integrated ERP systems for vendor-managed inventories. For example, direct integration of ERP systems was reported at Colgate:

> Colgate's plan is to use its network to get a peek at customer's stockpiles, while allowing its supplier to look at Colgate's inventory as well. The company is even supplying its most critical suppliers with computers loaded with R/3 system and plugged directly into the Colgate system. (Brownlee 1996, p. R17)

Unfortunately, directly linking systems is not simply a matter of allowing transactions to flow from one source to another. Instead, for different systems to be truly integrated there is a need for those systems to have some of the same MAPs, at least for the specific transactions being addressed. Otherwise, additional integration would be required.

ERP and Build to Order

Historically, firms have employed a build-to-forecast (BTF) approach to meeting demand. The BTF approach is first to forecast demand and then build product and resulting inventory to meet the demand. However, for products (e.g. computers) that change rapidly, such an approach can quickly lead to an outdated inventory. As a result, there has been an increasing focus on a new strategy – referred to as *build to order* (BTO) – particularly in the high-technology industries, where products rapidly become dated. With BTO, products are not built until they are needed. Using this approach, organizations maximize flexibility by keeping virtually all their inventory in raw materials.

Build to order replaces inventory with information systems technology (Teresko 1998). The transition from BTF to BTO requires an enterprise vision of information technology and processes. As noted by a vice-president of Compaq, "[w]e chose SAP AG's R/3 enterprise resource planning software to have a standard set of business rules to optimize decisions centrally for local execution" (Teresko 1998).

A critical part of BTO is the configuration engine, which provides the detailed inventory items in the product being built. For example, if a computer is being built then the configuration engine would detail the processes involved and the components (e.g., memory, hard drive) to be used. As noted by Trommer (1998), "[t]he new configuration model links to Compaq's SAP model with information on capacity to build and components on hand." The on-line real-time capability of ERP is necessary to provide the information needed for BTO.

At some point we can anticipate ERP-to-ERP communication for BTO, as described by a vice-president at Dell (Wagner 1999).

> Now picture an integrated world, where essentially the ... customer's ERP system automatically creates an order. It is by definition correct. The order can't be technically incorrect because the systems are talking to each other – there's no human element. The order goes straight down to the production line and, potentially 20 seconds later, the machine starts getting built, so you've eliminated a terrific amount of cycle time.

> On the back end of that, the moment the machine is finished being built, it gets shipped to the customer. The invoices get electronically transmitted right back into the customer's system, so the credit collection period starts immediately. The only limit is how long it takes to physically build the machine.

ERP and Electronic Commerce Links with Resellers

Enterprise resource planning systems also can be used to facilitate electronic linkages with supply chain partners – for example, resellers. Cisco does not stock finished goods, and most of its sales are by resellers. Although most of those resellers have relationships with distributors, they prefer to make their orders directly with Cisco on the Cisco website. As a result, it is important to try to establish an electronic linkage between resellers and their distributors. Cisco has made such a linkage through their website-connecting ERP systems.

As noted by Riggs (1999), "[r]ather than forwarding resellers' orders directly to the plants, orders can be routed to the distributor's ERP system. That way a distributor can add features, such as offering the reseller a discount if it purchases a certain number of products."

ERP and Merge in Transit

One of the more intriguing initiatives that has emerged as part of electronic commerce is that of "merge in transit" (MIT), which may be defined as a "service [that] collects shipments from multiple origin points and consolidates them, in transit, into a single delivery to the customer" (Dawe 1997).

Merge in transit is an issue either of merging and forwarding or of synchronization. In the former, separate components of an order can be stored in separate locations and merged on delivery; in the latter, multiple items are timed to arrive simultaneously. Merge in transit can have a substantial impact on a firm's inventory. Rather than having all component parts shipped to some central location and then merged and resent to customers when ordered, parts can be shipped directly from manufacturing sites to customers. Merge in transit makes heavy use of EDI (see e.g. Cooke 1998) and requires real-time information that could be acquired from an ERP.

As an example of MIT (Anonymous 1998), Dell Computer outsources manufacturing of the monitors it sells to a location in central Texas. In order to reduce its inventory and shipping expenditures, Dell has the manufacturer send the monitors to various UPS locations throughout the country. Monitors are timed to arrive at the UPS facilities at the same time as the computers. Customers receive a single consolidated shipment. As a result, Dell never inventories the monitors on its shelves.

Implementation of MIT is a relatively recent phenomenon in some industries. For example, on September 28, 1998, Cisco announced the following in their "Networked Commerce News":

> *New Consolidated Delivery Solution is Available*
> The Cisco consolidated delivery solution Merge in Transit (MiT) is now available to all U.S. customers. MiT is a new Internetworking Product Center (IPC) delivery option designed to simplify the receiving and remittance processes. With MiT, all items from a purchase order ship at the same time, and a single remittance invoice is generated. This eliminates multiple shipments and invoices, making it even easier for you to do business with Cisco. To select the MiT solution, simply choose the "Merge Order" delivery option in "Step 6, Taxes and Shipping" of the IPC order checklist.

As seen in this example and in the foregoing description, MIT makes a number of requirements of an ERP system. For example, although the products are emerging from multiple locations, there is a single purchase order and a single invoice, rather than invoices coming from each of the product locations. The ERP system must be able to account for sales of goods that are not ever "received" for sale to customers. Different locations communicate critical information using EDI.

Summary

Enterprise resource planning systems, network communications using EDI and WFT, and facilitating user access are the three building blocks of electronic commerce. These building blocks are not independent but instead are tightly linked for successful electronic commerce. The ERP system provides the information, communications link multiple enterprises, and widespread access is accomplished using the Internet and browser interfaces.

We have examined the role of ERP in a number of major electronic commerce initiatives. Enterprise resource planning was found to play important roles in facilitating customer tracing of goods, customer ordering of goods, vendor-managed inventories, building goods to order, linking with resellers, and merging goods in transit. These initiatives have led to a major reengineering of processes underlying electronic commerce. Throughout, ERP provides the central repository of information – an infrastructure and a technology that allows sweeping changes in processes that support electronic commerce.

References

Anonymous (1998). "Order Assembly Centers: Guarantee Same-Day Shipment in Any Quality." *Modern Materials Handling,* May.

Booker, E. (1999). "Web to ERP – ERP Stage II: Outsiders Invited In." *Internet Week,* October 25, ⟨www.techweb.com⟩.

Brownlee, L. (1996). "Overhaul." *Wall Street Journal,* November 18, pp. R12, R17.

Cooke, J. (1998). "Custom Built with Speed." *Logistics Management and Distribution Report,* January, pp. 54–7.

Dawe, R. (1997). "Move It Fast, Eliminate Steps." *Transportation and Distribution,* September.

Karpinski, R. (1999). "EDI Developer Extends E-Trading Boundaries." *Internet Week,* May 3, p. 10.

Kersnar, J. (1999). "The Tangled Web We Weave." *CFO Europe,* July/August, pp. 37–44.

McKenney, J., and Clark, T. (1995). "Procter & Gamble: Improving Consumer Value through Process Redesign." Report no. 9-195-126, Harvard Business School, Cambridge, MA.

Messmer, E. (1999). "Secrets of the E-Commerce Stars." May 25, ⟨www.cnn.com⟩.

Ricadela, A. (1999). "Cisco, 3Com Aim at ERP." *iweek,* August 2, ⟨www.iweek.com⟩.

Riggs, B. (1999). "Cisco Simplifies Business." *Information Week,* December 13, p. 92.

Roberts, B. (1998). "E-Commerce Poster Child Grows Up." *Datamation,* August, ⟨www. datamation.com⟩.

Teresko, J. (1998). "Replacing Inventory with IT." *Industry Week,* April 6, ⟨www. industryweek.com⟩.

Trommer, D. (1998). "BTO: New Order of Business." *Electronic Buyer's News,* November 16.

Wagner, M. (1999). "Next Step: Remove All Human Intervention." *Internet Week,* October 25.

Appendix 14-1

Implementing J.D. Edwards OneWorld at the University of Southern California's Leventhal School: An Interview with Professor Les Porter

Les Porter, Director of the Accounting Solutions Center, is on the faculty of the Leventhal School of Accounting in the Marshall School of Business at the University of Southern California (USC). Bob Kiddoo, also referenced in the interview, is on the faculty of the Leventhal School. Keck Center (Keck) is the computing and network support group of the Marshall School of Business.

What Is the School's Vision Regarding ERP Systems?

At the Leventhal School we have a vision that basically states that if our students are going to be valuable then they must have hands-on experience with the real thing. That is a vision that is now being taken up by many schools. Everybody is looking to get their students real-world experience, as real as possible.

As a result, we decided to integrate several enterprise packages into our curriculum. One of those packages was the J.D. Edwards client server package, OneWorld.

How Did You Implement It: Stand-Alone or Client Server?

When first implementing J.D. Edwards, we had two choices: implement OneWorld in the product demo version or in the full client server environment. Although some schools have implemented J.D. Edwards in the product demo version, we rejected this alternative from the beginning. It would have meant that each user must use the same machine in the lab whenever they are doing their exercises because their data is sitting on that machine. Since the machine is in the lab it is exposed to other students who could do damage to the machine or the data. If something happens, they are going to lose their data or access to the machine for some period of time.

When you think about the problems with the logistics where the students can only go in the lab and work on one machine, we decided to go straight to running it on client server, bypassing these difficulties. Ultimately, we were the third organization in the world to have J.D. Edwards OneWorld running in a client server environment.

What Unique Requirements Were Made by J.D. Edwards OneWorld Software?

J.D. Edwards is a just-in-time type of software that requires Windows NT in order to give the students access to the tool set. It downloads the modules that the user needs as they need them. J.D. Edwards uses a deployment server that keeps track of the software on a user's PC. The server goes out to the client to see if the client has the most current version, or any version of the software. If the user does not have the most current version or has no version at all, then the server will download it to the client.

This has a number of strengths and at least one weakness. First, you don't clutter up people's PCs with software that they don't need. Second, you only have to install the software on the deployment server and the system takes care of itself. As a result, there is no need to go around and update all the clients. Unfortunately, the problem for us is that when we use this in a class setting, our students touch all the modules. Each student ends up downloading everything. Thus, students don't just have a fat client, they have an obese one. Consequently, we ended up putting out some obese specialized NT clients in the labs.

What Problems Did You Have with the Client Server Version?

We had three problems, the first two of which are probably unique to the academic world. First, at USC we have multiple layers of security, with no central authority with control of all of it. Over the years, the system architecture has built up layer upon layer of security. As a result, in order to run the tool sets for all of the software we had to find a solution to our security problem. The support people at Keck generated a solution to our problem and continue to handle security.

Second, we had a problem with ODBC[1] drivers. ODBC drivers basically let one application look at the data in another application. In J.D. Edwards, the requirements for ODBC drivers are static. In a closed environment, static ODBC drivers may not be a problem. However, in an academic environment,

---

[1] ODBC denotes "open database connectivity," which is an API (application programming interface) for database access that uses SQL (structured query language) as its database access language. ODBC is based on CLI (call level interface) specifications from X/Open and ISO/IEC database APIs.

that becomes a nightmare. Every time that the system is upgraded somewhere, the ODBC drivers change and there is a conflict. We got to be very familiar with the ODBC drivers. Bob helped write some of the support information for J.D. Edwards regarding the ODBC drivers.

Third, we were lucky that we had the Parker & Parker lab money to get those special obese clients. However, we only were able to put out two machines that were powerful enough for the students. The data was up on the server so it did not matter which one, but it still had to be one of the two obese clients. Further, students had limited access because they could get to the machines only when the library was open. So in our environment, client server did not offer much of an advantage.

The Rest of the Vision

We are fully committed to giving the students 7 × 24 (seven days a week and 24 hours a day) access to any of the computing resources of the Leventhal School. Students operate in a significantly different part of the day than the school operates on. As a result, it was important to us to let the students work on this software during hours that were important to them, not hours that were convenient to the chief librarian. The only way we could do that was through the Internet. Unfortunately, it was not clear how.

Citrix Server

Last year I took some classes at the J.D. Edwards conference in Colorado. One class, not the presentation that I thought I was going to, was a Citrix presentation. This guy was demonstrating the software, using the Internet to connect to his office in Pennsylvania, that was hooking back to the lab in Denver. He was getting better response time than I had been getting in my hands-on class using the lab in the building. I got really excited about it – this was the greatest thing since sliced bread!

Citrix provides a window on a client that is running someplace else. We have a Citrix server that can run up to 20 clients on this one machine, based on our license. You are right back to the old days of time sharing. It can run 20 clients using the software so we don't care how obese the client is, since it is sitting right there in the lab. We can set it up and control it in the lab as Windows NT. We control the entire environment around it. We don't have to worry about client size: we have enough disk drives in there, it has 1 gig of memory and it has dual processors – it is our most expensive box. We don't have to worry about ODBC drivers getting changed.

It also is a huge step forward in managing your software. If you upgrade your software, you only have to change things in the server and this single client, and yet everybody sees the same software. We only need to worry about security for the client and the server. We don't need to worry about the ODBC drivers.

It gives the students access 24 hours a day from anywhere in the world from any box they have. They can be running Linux, Mac, or Windows 3.1 on an Intel 486, as long as they have a web browser and Internet access. They don't need Windows NT.

However, the truth is that it works best with Microsoft's Internet Explorer with a Citrix plug-in. We have had no problems with anyone using the most current release of Explorer, but we have had some problems with some of the older releases. It is the slickest thing I have ever seen for providing services to students.

It saved our bacon this year. As you know, we just moved into a new building for classes, Popovich Hall. Since it was brand new, the network applications were not available when classes started in the fall. We were faced with the question: "How could we get access to applications for classes?" We just put everything on Citrix. While others were struggling, all we needed was Internet access to get to the Citrix server.

I don't think that there is another school in the country with a Citrix server for their applications.

How Does It Work – Can You Show Me?

In order to get to the Citrix server and the different applications, I go to My-HomePage, which is just a local HTML file (Figure 14.2) – it could come up on a web page. Clicking on "Citrix" brings me to the Windows NT server logon for the Citrix server (Figure 14.3). At the terminal server level there are a number of programs that I can access, such as ACL for Windows, Great Plains, MAS 90, New Visio, and OneWorld (Figure 14.4). Clicking on OneWorld, I get the logon window for the OneWorld software (Figure 14.5), which leads directly to the software (Figure 14.6).

What Made It Work?

One of the advantages that we have is that the "Big 5" professional services firms and the Leventhal money have made it possible for us to get what we need. It is not an issue of do we do it, but how fast can we do it. Money was a big issue. We bought the box for the Citrix server. All total it was around $2,000 per user across the 20 user licenses, including the hardware and software, independent of the application.

But it was not just money, it was people. Mitchell Simons at Arthur Andersen deserves much of the credit for making J.D. Edwards work. It was the

Figure 14.2

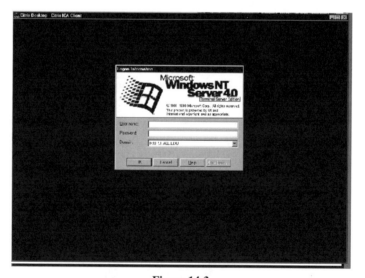

Figure 14.3

ability to pick up the phone and call Susan Karacostas of Arthur Andersen and get technical support. It was their technical people that spent a lot of time dealing with the J.D. Edwards people and got the system running. Our life would have been impossible without their support. The people over in Keck Center

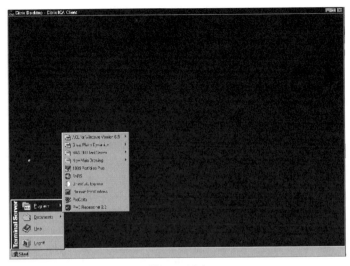

Figure 14.4

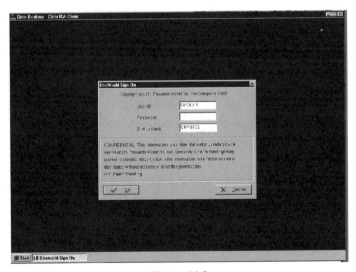

Figure 14.5

put a huge effort into getting our Citrix running. In fact, Keck continues to support and control the Citrix server.

Bob and I are not technical people. However, both Bob and I will take credit for what worked and blame the other for what did not work. Though the funny story is that when Bob was asked why USC had it up and running when no one

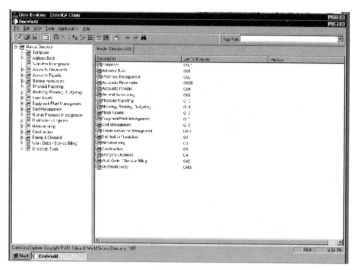

Figure 14.6

else did, he responded, "When Les told me to have it up and running in the class, I did not realize that there was an alternative." We just had to stay on it until it was done.

What Do You Plan for the Immediate Future?

Every application needed by students at the Leventhal School is up on the Citrix server, with 7 × 24 access: J.D. Edwards, Great Plains, MAS 90, Platinum, ACL, Pro Carta, and other software. In addition, Citrix also made it possible for students to get access to databases on CDs. We are already talking about putting in a second Citrix server, with automatic rollover. That way, if you lose one server, you have not lost all your capacity.

15

ERP Risk

Success and Failure Factors

Not all ERP implementations are successful. Implementations succeed and fail for a number of different reasons. Although we have pointed to a number of elements of risk or failure in ERP implementations, the purpose of this chapter is to provide a framework to facilitate risk identification and to identify some additional risks that can lead to ERP success or failure.

A risk is something that can go wrong. Risk is an exposure that can be a success factor if properly handled and a failure factor otherwise. Success factors and failure factors are two sides of the same risk. As in other settings, there is a trade-off between risk and return. Exposing the firm to risk may ultimately provide a greater benefit. For example, as we shall discuss, linking ERP to other applications increases the risk but also increases the potential return from the project.

One model for categorizing risk in the ERP system is given in Figure 15.1. In this matrix, risk is categorized along two dimensions: location in the ERP life cycle and type of application (technical, business, or organizational).

Types of Risk

Risks occur throughout the ERP system life cycle, which ranges from the go–no-go decision on ERP until after the system has gone live, including training issues. The types of risks and the extent of their impact on the organization vary as we move through the life cycle.

Technical risk refers to the risks that arise largely because of information processing technology. For example, technology used in ERP includes the operating system, relational database, client server technology, network requirements, and software. Business risks are those that derive from the models, artifacts, and processes that are chosen and adopted as part of the ERP implementation. Business risks are generated from the firm's portfolio of MAPs with respect

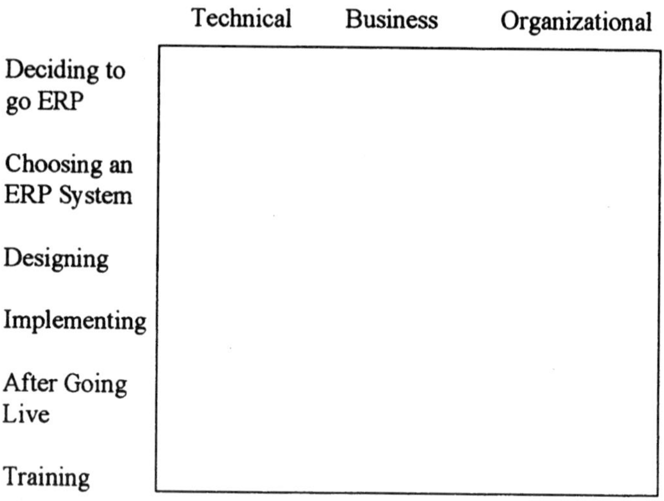

	Technical	Business	Organizational
Deciding to go ERP			
Choosing an ERP System			
Designing			
Implementing			
After Going Live			
Training			

Figure 15.1. Risk Matrix

to their internal consistentcy and their external match with business partners. Organizational risks derive from the environment – including personnel and organizational structure – in which the system is chosen and implemented.

In this discussion, risks will be placed in different categories, which may not be unique.

Risk Survey

A recent survey (Austin and Cotteleer 1999) analyzed the perceived risk of ERP projects across the three types of risk. The results, given in Table 15.1, establish that organizational risk is higher than business risk and that business risk is higher than technical risk.

Specifically, the survey reveals organizational risk as the most likely to be viewed as a "moderate to very high" risk. Similarly, higher-risk responses to business risk exceed those for technical risk. Although technical risk is important, it is dominated by both business and organizational risk. Problems brought on by technical risk are perhaps the easiest to fix, oftentimes just by upgrading (e.g., more server processing power).

General Technical Risks

Shifts in technology are necessary to employ ERP systems. As firms adopt ERP systems, they migrate to new technologies – different operating systems, relational database systems, client server computing, network systems – and,

Table 15.1. *Perceived Risk of ERP Projects (%)*

Level of Risk	Type of Risk		
	Technical	**Business**	**Organization**
Very low	10.5	4.5	1.5
Low	22.5	23.0	8.5
Moderate	39.5	32.5	18.5
High	15.0	26.0	37.5
Very high	11.5	14.5	35.0

Note: Based on averaging the results for 1998 and 1999 presented in Austin and Cotteleer (1999). Minor rounding differences prevent some columns from adding to 100.

accordingly, different processes. However, new technologies generate new risks. Many such risks are common to multiple portions of the life cycle; these are treated independently in this section.

Operating Systems

Operating systems have different capabilities and controls, and they require different knowledge. As a result, when firms implement ERP they need to choose an operating system and to employ people who understand that operating system. Initially, ERP systems were developed for and implemented in UNIX environments. However, when Microsoft implemented SAP in a Windows NT environment, it opened the door for other firms to do the same (Bashein, Markus, and Finley 1997). Until that time, the use of Windows NT had been limited to small implementations. Now the choice of operating system is another decision that can influence the knowledge and people required to implement and maintain the ERP system.

Client Server Computing

Client server is the dominant form of computing used for ERP. However, in many cases, legacy systems are embedded in mainframe computer environments. As a result, the firm's knowledge and experience may be based in an alternative form of computing, and hence there may be insufficient expertise and personnel for the new computing environment – particularly when client server is couched in some of the newer configurations, such as web access to time-shared client computing (see Appendix 14-1).

In addition, there can be different control structures in client server environments. In mainframe environments, virtually all the computing takes place on the mainframe. As a result, substantial effort is given to "bullet-proofing" the mainframe. For example, access is limited both physically and virtually, and there are likely to be backup computing facilities. However, in a client server environment, such access controls might be limited or not as fully implemented. Although the server and the mainframe should be subject to similar controls, a lack of knowledge or attention to the client server environment has sometimes left the server vulnerable. Further, in a client server environment, the shared computing between client and server means that the overall computing environment is more vulnerable and accessible. As a result, there are new risks associated with implementation of client server.

Network Capabilities

Enterprise resource planning systems are heavily dependent on the underlying network to provide wide-ranging user access by connecting dispersed client server systems, particularly when access is provided over the Internet, an intranet, or an extranet. In such settings, the network ultimately influences both the security and capacity of the system.

   Security of that network is critical to ensuring the overall security of the ERP system. Unauthorized network access can be the first step to unauthorized ERP access. *Capacity* of the network must be sufficient to allow access to the ERP system and any other applications supported by the network. Otherwise, the system will be too slow and not work as expected.

Database

Whereas legacy systems may have the same information stored in multiple databases, relational databases store information in only a single location. If something happens to that database then all the information could be lost. As a result, it is critical that there be appropriate back-up of the enterprise database.

Links to Other Systems and Independence of Computing Environment

Two other variables interact to influence technical risk. First, ERP systems may be established as stand-alone systems or they may be linked with other systems. Second, the ERP may have its own independent network and computing environment or it may share that environment with other applications. The interaction between these variables is summarized in Figure 15.2.

   Linking ERP to other systems increases the complexity and the resulting risk, in some cases leading to disastrous consequences. For example, one of the most publicized SAP implementation "failures" was that of FoxMeyer Drugs. That implementation was not just a SAP implementation. Instead, as noted by Jesitus (1997), it was an implementation of SAP and warehouse management software, the two of which required integration. In addition, at the same time the

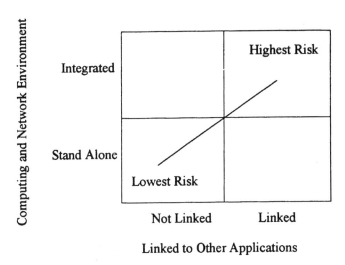

Figure 15.2. Environment and Application Linkage Interaction

primary warehouse was being shifted from a labor-intensive to a robotics-based warehouse. As a result, the engagement was enormously complex, fueling the ultimate engagement failure.

With regard to the second variable, if the system is introduced into a multiple application environment then many risks in those other systems will also be risks in the ERP system. If the ERP system has its own environment then its scope can be more easily controlled. Further, that environment can be customized to meet the specific needs of the ERP application.

Although risk is higher in the upper right quadrant, the potential for return is the highest there also. The ERP system provides its greatest benefit when integrated and linked to other applications.

Technical Risks and the ERP Life Cycle

Beyond those general technical risks, there are specific technical risks in each part of the ERP life cycle. These risks relate to computing, software, networks, and other technical issues.

Deciding Whether or Not to Do ERP

The decision of whether or not to do ERP may be influenced by the firm's current technical position. Those firms with substantial advanced technology capabilities are more likely to be in a position to understand the technical risks associated with ERP. Those firms that have not kept up with technology may find themselves unable to assess the technical risks, and may have to initiate their ERP efforts by first gaining such an understanding.

Technology risks can be minimized by seeing what has worked in the past and using similar configurations. As noted in Stein (1997a), "[i]t is people who make it harder for themselves by picking the wrong technology or by trying to push the envelope."

Choosing an ERP System

Choosing the ERP system requires hardware choice as well as software choice. The corresponding risks are likely to vary from company to company.

Typically, the primary concern is that the hardware be able to meet processing requirements. For vertically integrated companies, choosing the internally available hardware is a way of minimizing some types of risk. For example, Siemens Power Company (Hirt and Swanson 1998) initially considered Hewlett-Packard (HP) and IBM for their hardware choice. However, they ultimately chose hardware from Pyramid, a Siemens subsidiary, when Pyramid matched HP's and IBM's bid and also indicated that they would fly parts in from Germany within 24 hours. This minimized the risk associated with choosing vendors outside the Siemens family of companies.

It is inevitable that no software will be optimal for everyone. Virtually all software choice is, in part, a political process, and any numerical weighting system can be manipulated. Further, requirements change. As a result, there always is a risk of choosing software that does not continue to meet the needs of the firm.

ERP Design, ERP Implementation, and Going Live

Technical risks from system design, implementation, and going live stem from a number of sources, including computing and network issues.

In a client server computing environment, as with all computing environments, it is possible to underestimate computing requirements – particularly if access is given to the system via an extranet. Cisco's initial implementation did not have enough computing capacity (Cotteleer, Austin, and Nolan 1998). Fortunately for Cisco, they had purchased the hardware on a promised capability rather than a specific configuration, so the vendor was required to fix the problem.

The capacity of the network also can determine the overall success of the system. A discussion with a PricewaterhouseCoopers consultant illustrates the extent to which the network can constrain the system success. In one case, a test of the network capacity yielded the finding that each transaction required ten seconds from initiation at the client to capture at the server. A quick computation reveals that such a network structure could process only six transactions per minute: 360 per hour or roughly 3,600 for a ten-hour day. For a large firm with literally millions of transactions (orders, etc.) per year, such an infrastructure is insufficient. As a result, there needs to be a matching process between communications capabilities and processing requirements.

Training and Personnel

Throughout the firm, new technologies require that employees gain new knowledge about how to implement and use that technology and how that technology facilitates ERP system usage. Either existing employees can be given the appropriate training or new employees can be sought out. In any case, the ERP system may require employees with different capabilities and skills, forcing a change in personnel. Any major change in personnel is itself a risk concern.

Business Risks

Business risks derive from (1) the firm's choice of models, artifacts, and processes, (2) how well those MAPs work in the organization, and (3) how well they facilitate interaction with other partner firms (e.g., vendors and suppliers). In addition, business risks relate to how well the system allows the company to compete.

Deciding Whether or Not to Do ERP

Insufficient resources is one of the biggest risks affecting the decision of whether or not to do ERP, and the rationales used for making this decision can also result in business risk. Not implementing ERP means forgoing an infrastructure that could be used to integrate across the supply chain and so maintain competitiveness.

Having sufficient resources is one of the most important requirements. Without sufficient resources, ERP creates huge business risk. Enterprise resource planning systems can require a large expenditure of resources – typically millions and sometimes hundreds of millions of dollars. For example, FoxMeyer planned $65 million for their SAP implementation. However, by choosing to do ERP, firms may be committing to an expenditure of resources that is not appropriate. Firms implementing an ERP may go bankrupt either because (in part) of the ERP or completely independent of the implementation. As an example of the former, FoxMeyer claimed in litigation that SAP was one of the reasons that it had gone bankrupt, and it is now suing both SAP and Andersen Consulting (see Radosevich 1998). Ultimately, SAP was implemented by the firm that took over FoxMeyer. As an example of the latter case, Geneva Steel declared bankruptcy the day after their $8 million SAP system was implemented.

As discussed in Chapter 7, technology rationales can provide a basis for computing costs and benefits of going ERP. However, those rationales provide no basis for making decisions about MAPs. As a result, firms need to focus on which processes need to change (and how they need to change) early in the ERP life cycle. Those process changes can then be used to guide choice of MAPs and to evaluate the implementation. Otherwise, the firm risks having no basis for choosing their MAPs.

Making the decision to go with ERP can be critical to future partnerships. For example, increasingly firms are integrating across the supply chain via their ERP systems. As a result, the choice to go ERP may influence the ability of the firm to be a part of emerging supply-chain partnerships. Not going with ERP can thus cause a firm to lose out on opportunities.

Choosing an ERP System

ERP system choices often are mandated by strategy or based on responding to the competition. If a firm's primary competitors have implemented an ERP system, then the biggest risk may be not choosing to implement an ERP. In such a setting, the competitor may be in a position to offer improved customer services through such capabilities as "available to promise," as seen at Quantum Computer. In those settings the firm is likely to risk facing competitive disadvantage or loss of "first mover" advantage.

In some cases, system characteristics can influence whether or not a particular ERP system will be successful. For example, many legacy systems have been specifically designed to meet high–transaction volume needs. If the new system cannot handle the volume requirements placed upon it, then it can fail. Firms such as FoxMeyer have reportedly gone into bankruptcy because of the inability of the ERP system to handle the volume of input required. For example, Radosevich (1998) indicated that one of the reasons cited by FoxMeyer for the failure of the ERP engagement was that the resulting system could process only 10,000 invoice line items per night, although the requirements were 420,000.

ERP Design

Perhaps the most important issue in design is that the firm choose an ERP design that provides the most favorable cost–benefit relationship. What are some potential causes of suboptimal design? In the first place, design – like choice – is a political process. In the design process there are winners and losers, and these fates are often the outcome of a political process. Second, in some situations project team members make design choices that should have been made by process owners or management. As a result, processes don't work as expected or do not mesh well with other processes. Third, costs and benefits may be difficult or impossible to determine accurately, making it hard to discriminate among alternatives. Fourth, at the design stage, people begin to see how much staffing and jobs will change and may become concerned about those changes. For example, one project was stopped because "[t]his project would have changed how people work and reduced staffing by half. It was the easiest thing to cut because people did not have the stomach for it" (Stein 1997b).

ERP Implementation

Two of the biggest business risks are that the implementation will take longer and cost more than expected. For example, Hershey officials noted that their SAP–Siebel–Manugistics implementation was three months behind schedule, offering this as partial explanation for why expected earnings were 10% off (Branch 1999). Because the system was delayed, order taking and distribution were disrupted. A U.S. Government site blamed its failed SAP implementation on a number of factors, including cost overruns (Stein 1997b).

After Going Live

After going live, if the system is not working properly then there can be problems with both customers and suppliers. For example, as noted in Koch (1999), "[c]ustomers did not get deliveries, or they got the wrong amount or the wrong products, you name it." Firms need to overcome such problems, which are characteristic of the stabilization period, as rapidly as possible. Otherwise, the firm faces potential loss of customers and vendors or increased costs of expediting late orders.

Training

Training should provide users with process and system information. Although training is frequently underestimated, it can entail a number of other business risks as well (too early, too late, not enough detail, etc.); see Chapter 13.

Organizational Risks

Organizational risks derive from the people, organization structure, and environment in which the system is chosen and implemented.

Deciding Whether or Not to Do ERP

Perhaps the biggest risk is associated with the decision as to whether or not to do ERP. One of the most important organizational variables is to ensure top management's involvement. A number of researchers (e.g. King and Konsynski 1993) have noted the importance and advantages associated with top management's buying into the decision, if not making the decision. Without top management's interest there may not be sufficient resources for the engagement nor the authorization necessary to change processes. Reportedly, one of the primary factors leading to a failed SAP engagement in the federal government of the United States was "a lack of executive support" (Stein 1997b). Similarly, Perrier's SAP implementation also suffered from a lack of top management support (Stein 1997a).

In addition, with ERP there is a need to have buy-in from the domain and not only from top information services management. Domain managers need to buy in because of the large potential for process change. For example, as seen in the Microsoft case, Microsoft had two failed ERP implementations. It was not until finance top management bought in to ERP that it was a success.

Even so, a lack of top management support does not always result in failure. For example, Siemens Power apparently did not have top management support (Hirt and Swanson 1998). However, that engagement was completed in a timely and cost-effective manner and the system is now heavily used.

Choosing an ERP System

This stage in the ERP life cycle includes some "people" issues – for example, choosing the right consultant. Although a number of variables can influence choice of consultant, perhaps the dominant choice variable is the quality of previous, similar consulting engagements. However, previous engagement is not the only variable, as seen in the case of Siemens Power (Hirt and Swanson 1998). At Siemens, the ERP project team solicited bids from IBM, HP, and Siemens Nixdorf (SNI, another Siemens company). As noted by a member of the Siemens computer information systems (CIS) management,

> I did not think that we would get a competitive bid from any one of the Big 5 companies. They would not bother with us .... But with IBM we had a good relationship and so we did get a bid from them. We turned them down because I felt they had nobody on the team that would be here that had manufacturing experience that I could see ... and they were more expensive than SNI. So we finally went with SNI.

Siemens Power chose the Siemens company, again minimizing the risk associated with going outside the Siemens family.

ERP Design and Implementation

Enterprise resource planning requires that models of the organization be built into the software. Those ERP models influence the flow of information and can therefore change the actualized organization design. For example, being forced to make each department either a profit center or a cost center (Bashein et al. 1997) means that information for evaluating costs and profits would now be available, where none was before. A change in available information can change organization relationships.

In addition, as noted in the discussion about big-R reengineering, firms often change the software to meet their needs. However, this can be costly and increase the engagement time. For example, Automated Packaging Systems (APS) abandoned their ERP system because it was a year late. As noted in Zerega (1998), "[t]he delay stemmed from the large amount of development that the ERP vendor needed in order to fit its application to APS's work processes."

We observed previously that small-r reengineering frequently involves changing the firm's processes to conform to the software. This, in turn, can lead to changes in the organization and resultant shifts of power within an organization. In order to accommodate its ERP system, one firm changed its compensation system to one based on salary. Such changes generally are unanticipated and difficult to identify before the implementation.

After Going Live

Cultural issues are among the most important organization risks associated with going live. An ERP system can be rejected if it does not fit the culture. Enterprise resource planning technology alone does not change the culture; management also needs to make substantial efforts. For example, forum discussant Thomas Davenport noted that

> firms assume that SAP – more or less by itself – will change the culture of the organization. If SAP, the company, communicated strongly that no software package can change a culture without concerted efforts from managers, you'd probably have a higher implementation success rate. Maybe the company does that already, but a lot of its customers aren't getting the message. (CIO 1996)

Training and Personnel

As noted earlier, ERP systems push data input closer to the source of the information. In some cases, this means that employees not accustomed to data input will take on the task. For example, an ERP may shift the task of data input out to the loading dock personnel. But if they don't know how to use the system, it can fail. Hence, particular care must be taken to ensure that the employees are given sufficient education to enable proper use of the system.

However, the primary organization risk in training and personnel, is that there may not be adequately trained people *after* the system has been implemented. For example, as noted in Ferranti (1998), "[w]hen projects fail,... a common reason is that key project staff are hired away by the company's competition, leaving people unfamiliar with the implementation to manage it." As a result, there need to be appropriate opportunities available for the project team to continue, after implementation, within the company.

Summary

This chapter established a model using type of risk (technical, business, or organizational) as one dimension and the ERP life cycle as the other dimension. Analysis of a survey found that the greatest risk was organizational, followed by business risks and then by technical risks. The remainder of the chapter elicited some of the risks in each of those categories, ranging across the entire life cycle.

References

Austin, R., and Cotteleer, M. (1999). "Current Issues in IT: Enterprise Resource Planning." Unpublished presentation, October.

Bashein, B., Markus, L., and Finley, J. (1997). *Safety Nets: Secrets of Effective Information Technology Controls.* Morristown, NJ: Financial Executives Research Foundation.

Branch, S. (1999). "Hershey to Miss Earnings Estimates by as Much as 10%." *Wall Street Journal,* September 14, p. B12.

CIO (1996). "SAP Key Roles." ⟨www.cio.com/forums/061596_sap_roles.html⟩.

Cotteleer, M., Austin, R., and Nolan, R. (1998). "Cisco Systems, Inc.: Implementing ERP." Report no. 9-699-022, Harvard Business School, Cambridge, MA.

Ferranti, M. (1998). "Debunking ERP Misconceptions." *InfoWorld,* August 17, p. 46.

Hirt, S., and Swanson, E. B. (1998). "Adopting SAP at Siemens Power Company." Paper presented at the International Conference on Information Systems (Helsinki), December.

Jesitus, J. (1997). "Broken Promises." *Industry Week,* November, ⟨www.industryweek.com⟩.

King, J., and Konsynski, B. (1993). "Singapore TradeNet (A): A Tale of One City." Report no. 9-191-009, Harvard Business School, Cambridge, MA.

Koch, C. (1999). "The Most Important Team in History." *CIO Magazine,* October 15.

Radosevich, L. (1998). "Bankrupt Drug Company Sues SAP." *InfoWorld,* August 27, ⟨www.infoworld.com⟩.

Stein, T. (1997a). "IT Stalls at Xerox and Perrier." *Information Week,* March 31, p. 20.

Stein, T. (1997b). "Federal Site Drops SAP." *Information Week,* December 22, p. 36.

Zerega, B. (1998). "Customer Information Center Supports Front Line." *InfoWorld,* August 10, p. 58.

Index